Use of Microcomputers in Geology

COMPUTER APPLICATIONS IN THE EARTH SCIENCES
A series edited by Daniel F. Merriam

1969-Computer Applications in the Earth Sciences

1970-Geostatistics

1972-Mathematical Models of Sedimentary Processes

1981-Computer Applications in the Earth Sciences: An Update of the 70s

1988-Current Trends in Geomathematics

1992-Use of Microcomputers in Geology

Use of Microcomputers in Geology

Edited by

Hans Kürzl
Logistic-Management-Service
Leoben, Austria

and

Daniel F. Merriam
Kansas Geological Survey
University of Kansas
Lawrence, Kansas

Springer Science+Business Media, LLC

Library of Congress Cataloging in Publication Data

Use of microcomputers in geology / edited by Hans Kürzl and Daniel F. Merriam.
p. cm.—(Computer applications in the earth sciences)
"Based on proceedings of an international geological congress symposium and poster session on the use of microcomputers in geology, held July 1989, in Washington, D.C."—T.p. verso.
Includes bibliographical references and index.
ISBN 978-1-4899-2337-0
1. Geology—Data processing—Congresses. 2. Microcomputers—Congresses. I. Kürzl, Hans. II. Merriam, Daniel Francis. III. Series.
QE48.8.U84 1992 92-29735
550.'.285'416—dc20 CIP

Based on proceedings of an International Geological Congress symposium and poster session on the Use of Microcomputers in Geology, held July 1989, in Washington, D.C.

ISBN 978-1-4899-2337-0 ISBN 978-1-4899-2335-6 (eBook)
DOI 10.1007/978-1-4899-2335-6

Originally published by Plenum Press, New York in 1992
Softcover reprint of the hardcover 1st edition 1992

CONTRIBUTORS

Agterberg, F.P., Geological Survey of Canada, 601 Booth Street, Ottawa, Ontario, K1A 0E8, Canada

Brower, J.C., Department of Geology, Syracuse University, Syracuse, New York 13244, USA

Buccianti, G.F., Ecosystems s.a.s., Via Mariti 10, 50127 Firenze, Italy

Detay, M., International Training Center for Water Recources Management, B.P. 113, 06561 Valbonne, France

Duncan, D.McP., Department of Mines, Tasmania, Australia

Dutter, R., Institute for Statistics and Probability Theory, Vienna, Austria

Fisher, N.I., CSIRO Division of Mathematics and Statistics, Sydney, Australia

Gelin, A., Macquarie University, Sydney, Australia

Gunther, F.J., Computer Science Corporation, 1100 West Street, Laurel, MD 20707, USA

Guth, P.L., Department of Oceanography, U.S. Naval Academy, Annapolis, Maryland 21402, USA

Holub, B.B., Institute of Geophysics, Mining University, Leoben University, A-8700 Leoben, Austria

Karnel, G., Institute for Statistics and Probability Theory, Vienna, Austria

Kürzl, H., Logistic-Management-Service, Franz-Josef-Straße 6, A-8700 Leoben, Austria

Merriam, D.F., Kansas Geological Survey, University of Kansas, Lawrence, Kansas 66047, USA

Minissale, A., Cnt. St. Mineral. Geochim. Sedim. (C.N.R.), Via La Pira 4, 50121 Firenze, Italy

Omoumi, H. Department of Geology, University of Alberta, Edmonton, Alberta, T6G 2E3, Canada

Pereira, H.Garcia, CVRM - Instituto Superior Técnico, Av.Rovisco Pais 1096, Lisboa Codex, Portugal

Powell, C.McA., Department of Geology, University of Western Australia, Perth, Australia

Poyet, P., Department of Computer Science, CSTB, B.P. 141, 06561 Valbonne, France

Smith, D.G.W., Department of Geology, University of Alberta, Edmonton, Alberta, T6G 2E3, Canada

Soares, A., CVRM - Instituto Superior Técnico, Av. Povisco Pais 1096, Lisboa Codex, Portugal

Sternberg, R.S., Department of Geology, Franklin and Marshall College, Lancaster, Pennsylvania 17604-3003, USA

Tabesh, E., Department of Geology, Syracuse University, Syracuse, New York 13244, USA

Vannier, M., Ecole des Mines Paris, 35 rue St Honore Fontainebleau, France

Woodtli, R., Geodidax, 21 Avenue du temple, CH-1012 Lausanne, Switzerland

PREFACE

This volume 'Use of Microcomputers in Geology' is the sixth in the series *Computer Applications in the Earth Sciences* published by Plenum Press in New York. The series was started in 1969 to publish proceedings of important meetings on geomathematics and computer applications. The first two volumes recorded proceedings of the Colloquia (1969, 1970) sponsored by the Kansas Geological Survey at The University of Kansas in Lawrence. The third volume was proceedings of the 8th International Sedimentological Congress (1971) held in Heidelberg, West Germany; the fourth was preceedings of the 8th Geochautauqua (1979) at Syracuse University in Syracuse, New York; and the fifth was selected papers from the 27th International Geological Congress (1989) held in Washington, D.C. All meetings were cosponsored by the International Association for Mathematical Geology.

These special publications are important in the development of quantitative geology. Papers by a wide range of authors on a wide range of topics gives the reader a flavor for recent advances in the subject – in this volume, those advances in the use of microcomputers. The 24 authors of the 15 papers come from nine countries – Australia, Austria, Canada, France, Italy, Portugal, Switzerland, UK, and USA. My coeditor, Hans Kürzl, has given pertinent information on the included papers in the Introduction.

Microcomputers made their first impact in the earth sciences in about 1982. In nine years they have permeated every conceivable nitch in the science from workstations to laptop field computers. They are used for everything from number crunching, graphics to electric communication, to word processing. They have become a way of life. Here then in this volume is a collection of papers extolling some of the virtues of micros.

For the purist, it should be noted that all the papers in the IGC session on 'Microcomputer Applications in Geology' were not available for publication. Therefore, a few papers on the subject were 'commissioned' from practioners willing to make a contribution to 'fill out' the volume. It is sincerely hoped that this potpourri will inform and inspire those interested in micros.

Several people helped with the preparation of this volume. I want to thank the anonymous reviewers. Patricia M. Vann of Plenum Press arranged for publication of the volume. The authors are to be commended for their contributions, which if the volume enjoys any success will be because of their efforts.

Skara, Sweden
June, 1991

D. F. Merriam

CONTENTS

INTRODUCTION

The 28th International Geological Congress (IGC) took place in Washington, D.C. from 9 - 19 July 1989. A comprehensive poster session covering nearly all fields of geoscience accompanied the technical program. COGEODATA, the IUGS Commission on Storage, Automatic Processing and Retrieval of Geological Data and IAMG, the International Association for Mathematical Geology have been active over many years to increase the knowledge, the development, and training in computer applications for geoscientific data. Consequently this topic was included in the IGC's poster program under the title "Microcomputer Applications in Geology" (H.Kürzl and J.O.Kork, Convenors).

During my affiliation with "Joanneum Research Association" until 1989 the main activities of the Mineral Resources Research Division in Leoben were dedicated to the development of data analysis and presentation systems for the regional geochemical and geophysical surveys of Austria. At the beginning of these projects in the early eighties already contacts were made to COGEODATA members and representatives to get scientific support and insight into the latest developments and ideas in geodata-processing. Especially R.Sinding-Larsen, P.Leymarie, R.G.Garrett, and R.B.McCammon supported our new group with excellent advice. First results were presented on a joint seminar with COGEODATA in Leoben at the end of 1984.

The following years were characterized by intensive project work, many personal visits, scientific communications, and presentation of research results within COGEODATA/IAMG activities. The projects of the Austrian regional mineral reconnaissance surveys have been finished in the last few years by using latest techniques in data reduction and multivariate analysis, which have been assembled, modified, and completed to packages on our grown-up Leoben computer system.

The invitation for convening the microcomputer poster session by the IGC organizing committee in 1988 was a big surprise but a challenging mandate. It offered an interesting opportunity to shape this small but important event as part of the large extensive technical program. I really do hope, although I have changed my position now to a free consulting engineer, the friendly connection to all these people will continue and further common activities will arise in the future.

The call for papers led to the acceptance of 15 posters, with contributions from 12 different countries all over the world. The session was held with 12 presentations on Friday morning, 14 July 1989, and caused a lot of attention among the congress participants. Due to the actual political development in the first half of 1989 the three Chinese participants were unable to come and their three announced posters had to be withdrawn.

Encouraged by the chairmen of COGEODATA and IAMG a preliminary call for papers for publication was carried out at the Congress and ten participants documented their interest in preparing a manuscript. This was enough support to continue with that idea. The results can be seen on the following pages. Studying the contributions, we can recognize that microcomputers have already become an integral part in all fields of geoscience. The development is yet progressing and goes rapidly in highly sophisticated applications, such as artificial intelligence tools and techniques. But also in the practical fields, microcomputers have spread. They now support all types of surveys. They are used in professional resource developments as well and represent an integral part of many modern geoscientific teaching and training courses.

In the following let me introduce the authors and their papers presented:

H.GARCIA PEREIRA and A. SOARES (Zoneography of mineral resources) present a new multivariate data-analysis approach to detect continuous zones with similar characteristics in the early stage of mineral-resource evaluation. In combination with factorial analysis, graphical representation of groups, expert geological advice, and morphological geostatistics, a methodology to cope with zonation problems was developed and is illustrated by two case studies.

M.VANNIER and R.WOODTLI (Teaching and testing strategy in mineral exploration by simulation techniques on personal computers) give a detailed description of the evolution of use of microcomputers and related software in the field of geological modeling. It is shown how this was introduced in teaching mineral exploration and how it can be applied to simulate geological data training programs in geostatistics and multivariate geodata analysis.

PATRICE POYET (Computer-aided decision techniques for hydrogeochemical uranium exploration) introduces multivariate data-analysis techniques under the aspect of different geological data types and their relevant heterogeneous behavior. He describes a very efficient set of data processing, data interpretation, and data representation software available on microcomputers, offering assistance to the exploration geochemist especially when faced with decision-making problems.

F.P. AGTERBERG (Estimating the probability of occurrence of mineral deposits from multiple map patterns) discusses the problems of estimating the probability of the occurrence of mineral deposits from map patterns. This paper is a continuation of his investigation on the probability of occurrences modeled as a function of map-element characteristics. Weights of evidence modeling and logistic regression are discussed as methods for estimating the probability.

ROBERT S. STERNBERG (Use of a laptop computer and spreadsheet software for geophysical surveys) describes the successful introduction of laptop computers and spreadsheet software to many types of geophysical problems and shows applications to geophysical field work for data acquisition, logging, reduction, and plotting. Examples discussed range from seismic data acquisition to gravity data reduction and gravity modeling.

B.B. HOLUB (A program for petrophysical database management) gives details on the computer program ROCKBASE which manages data obtained in the laboratory from field samples. The petrophysical parameters measured on the samples can be from hydrogeology, geochemistry, or petrography.

PATRICE POYET and MICHEL DETAY (Artificial intelligence tools and techniques for water-resources assessment in Africa) present the development of a hydrogeological expert system able to handle the drilling location problem within the scope of village water-supply programs based on experience gained by the authors in fifteen African countries. The recognition of relevant hydrogeological parameters and examples of advanced computer knowledge modeling methods are described in detail.

ANGELO MINISSALE and G.F. BUCCIANTI (Hydrodat®: a package for hydrogeochemical data management) describe a PC-based database oriented system especially designed to handle and analyze hydrogeochemical data. It gives positive evidence that the integration of different commercially available software modules to one package can lead to efficient solutions for many different working fields in hydrogeochemistry.

D.G.W. SMITH and H. OMOUMI (MinIdent - some recent developments) describe the MinIdent program as adapted to a PC from the mainframe. MinIdent is a database containing some 1500 descriptions of mineral species and their synonyms which are used primarily for the identification and retrieval of minerals and associated data.

P.L. GUTH (Microcomputer application of digital elevation models and other gridded data sets for geologists) is concerned with the program MICRODEM which was developed for terrain analysis using a digital elevation model. The program, adapted to a PC, handles a variety of gridded digital data sets.

FRED J. GUNTHER (Reusable code works!) reports on some personal experience in microcomputer applications programming for statistical analyses and graphic displays. He shows that reusing code has increased software reliability as well as programmer productivity and has decreased development costs significantly.

N.I. FISHER, C. McA. POWELL, A. GELIN, and D. McP. DUNCAN (A PC statistical package for field acquisition and analysis of two-dimensional orientation data) present a PC statistical package for orientation data. The versatile package can be used in comparing and combining summary values from different data sets.

J.C. BROWER and D.F. MERRIAM (A simple method for the comparison of adjacent points on thematic maps) outline a simple method for comparing adjacent points on a series of maps. Similarities or differences can be computed and the correlation, Euclidean, or Mahlanobis coefficient plotted and contoured to depict these similarities and differences. The resultant patterns then can be related to geological features of the original data sets and further analyzed by other statistical techniques.

E. TABESH (Map integration model applied in site selection) introduces a concept of a map integeration model that has the decision-making ability to aid in the selection of the most suitable site. She takes into consideration both structured (well-defined) and unstructured (ill-defined) decision-making which includes the concept of fuzzy set theory.

R. DUTTER and G. KARNEL (Analysis of space and time dependent data on a PC using a data-analysis system (DAS): a case study) make a presentation on their data analysis system for the PC. The program is easy to use for exploratory, numerical, or graphical statistical data analysis and is flexible for use for just about any interactive graphical analysis.

The diffusion of micros in all fields of geoscience is obvious and it will become an essential tool covering the wide range from science to practice. However, we should never forget that they are a product of our mind and will hopefully never replace creativity and personal know-how in research and applied work.

I want to thank especially G. Gaal of COGEODATA, R. McCammon of IAMG, and J. Wolfbauer of Joanneum Research for their friendly encouragement to proceed with this project and their steady support and assistance. Special thanks are due to D. Merriam, who identified himself with the project spontaneously and offered the necessary cooperation in final editing and preparation of the book. In addition, appreciation is given to Dr. H. Baeck of Logistic-Mangement-Service Leoben, who made available his office facilities to do all the necessary preparation work for editing the manuscripts and for carrying out all the management activities to ensure a good and interesting publication.

Hans Kürzl

Logistic-Management-Service
Franz-Josef-Straße 6
A-8700 Leoben, Austria

ZONEOGRAPHY OF MINERAL RESOURCES

H.Garcia Pereira and A. Soares
CVRM - IST, Lisbon, Portugal

ABSTRACT

In facing the problem of the exploitation of heterogeneous mineral resources (ore or oil), the question of defining continuous zones exhibiting similar characteristics may arise. A method to approach this question using factorial and geostatistical techniques was developed. It contains three main steps:

(1) Taking as an input the data matrix (samples x variables), a factorial technique (principal components analysis or correspondence analysis) is applied, giving rise to groups of samples of similar characteristics.

(2) The problem of contiguity within samples of the same group and the number of groups to be retained is solved by expert advice of the geological/exploitation team, based on graphical representation of zones.

(3) Once decided which samples belong to each final group, the boundaries from zone to zone are estimated, using a transitive kriging technique, relying on the geometric variogram.

Use of Microcomputers in Geology, Edited by D.F. Merriam and H. Kürzl, Plenum Press, New York, 1992

Two case studies are presented to illustrate the method: The first one discusses a polymetallic sulfide orebody located in the South of Portugal, which is to be split into zones feeding different mineral-processing units. In the second, a Middle East petroleum reservoir is divided into homogeneous zones in order to improve the secondary oil-recovery planning.

The method presented here, combining factorial analysis and geostatistics, is a useful tool for the purpose of delineating zones in heterogeneous deposits of mineral resources. It provides estimates of boundaries between zones based on their geometric structure and gives a reliable basis for further exploitation planning.

INTRODUCTION

When planning the development of large mineral resources, the problem of defining continuous zones with similar characteristics may arise.

When the information available on each sample includes a great variety of attributes of different types, a data reduction procedure based on multivariate statistics is helpful for establishing the zonation guidelines. But the groups of samples provided by the classification procedure must exhibit spatial contiguity in order to meet the requirements of the exploitation method. Furthermore, it usually is necessary to estimate the morphology of each zone by the group of samples with similar properties, which constitutes a technological unit.

By combining factorial analysis, graphical representation of groups, and expert geological advice and morphological geostatistics, a methodology to cope with zonation problems was developed. It contains three main steps:

(1) Taking as an input the data matrix (samples x variables), a factorial technique (principal components analysis or correspondence analysis) is applied, giving rise to groups of samples of similar characteristics.

(2) The problem of contiguity within samples of the same group and the number of groups to be retained is solved by expert advice of the geological/exploitation team, based on graphical representation of zones.

(3) Once decided which samples belong to each final group, the boundaries from zone to zone are estimated, using a transitive kriging technique, relying on the geometric variogram.

Two case studies are presented to illustrate the proposed methodology: the first one regards a polymetallic sulphide orebody located in the south of Portugal, which is to be divided into zones feeding different mineral processing units. In the second, a Middle East petroleum reservoir is divided into homogeneous zones, in order to improve secondary oil-recovery planning.

A MASSIVE POLYMETALLIC SULPHIDE OREBODY

The available data were obtained from several deep drillholes, assayed on 1 m sections for 11 elements - Cu, Pb, Zn, S, Fe, Ag, Hg, Sn, As, Sb, and Bi. The element grades were arranged in matrix form. The corresponding rows represent the 780 samples.

In a first step, the data matrix of 780 samples with 11 variables was submitted to an algorithm performing the principal components analysis (PCA) of standardized data (Lebart, Morineau, and Warwick, 1984). The output gives projections of the samples and variables onto the main factorial axes.

Results of this procedure are shown in Figure 1, where projections of variables onto Axis 1 and 2 are displayed, as well as the initial limits defining 3 groups of samples in the plane of the two principal axes.

Limits on Figure 1 then were changed interactively, according to geological context, until a final morphological definition of zones is reached. The spatial continuity of each zone is visualized through the representation of experimental data, and its final shape is estimated via morphological kriging (Soares, 1990), as displayed for the pyritic ore type in Figure 2.

MIDDLE EAST OIL FIELD

Data on an extensively explored oil reservoir were taken from 172 wells. Their locations are given in Figure 3.

The set of variables available on each well was divided into two categories: the first includes the elevation of the oil-bearing formation and water saturation, whereas the second contains porosity, permeability, and facies. For the purpose of correspondence analysis application

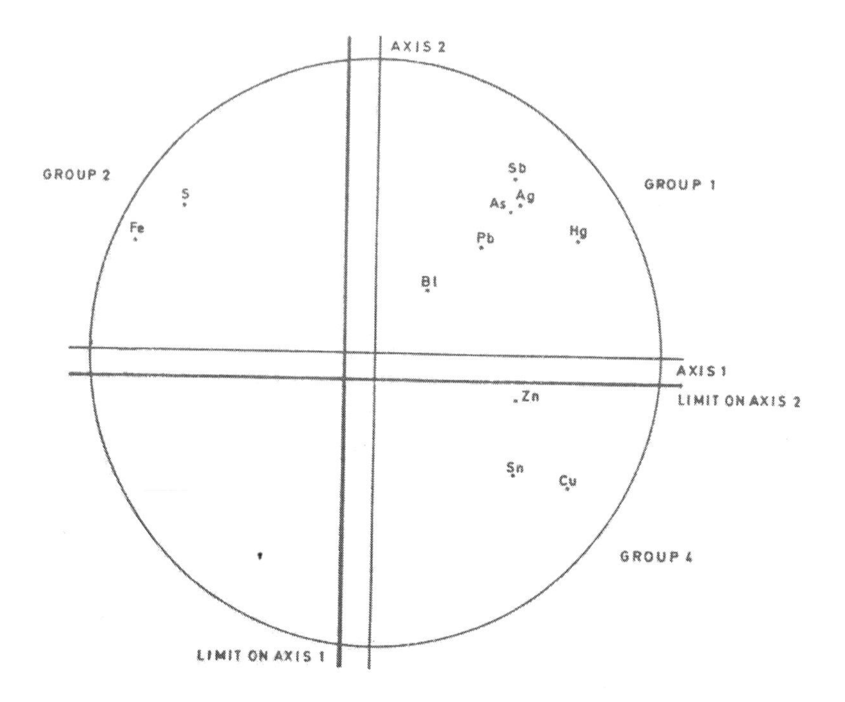

Figure 1. Results of PCA of standardized data and splitting procedure

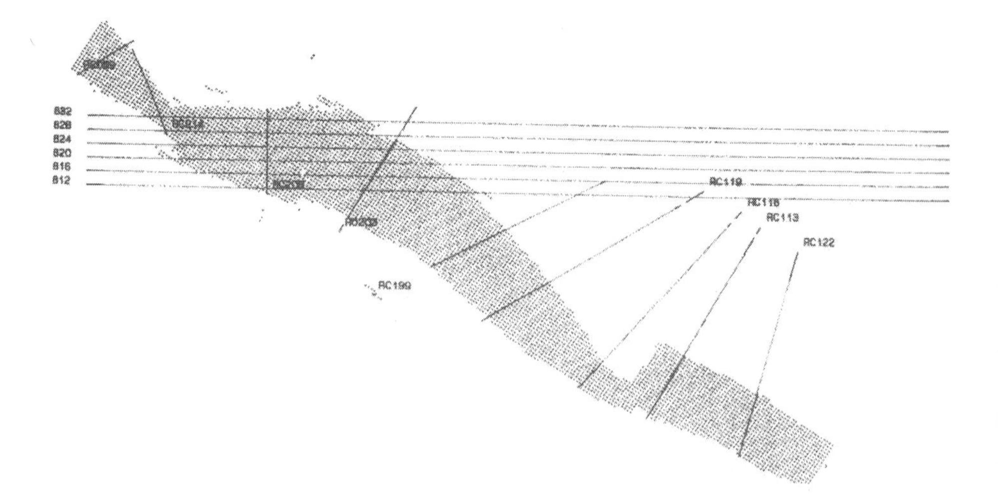

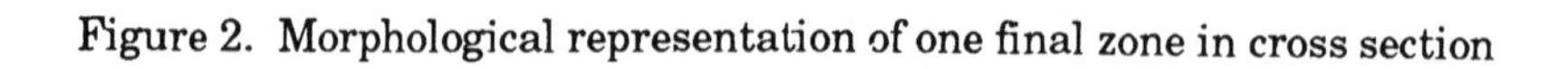

Figure 2. Morphological representation of one final zone in cross section

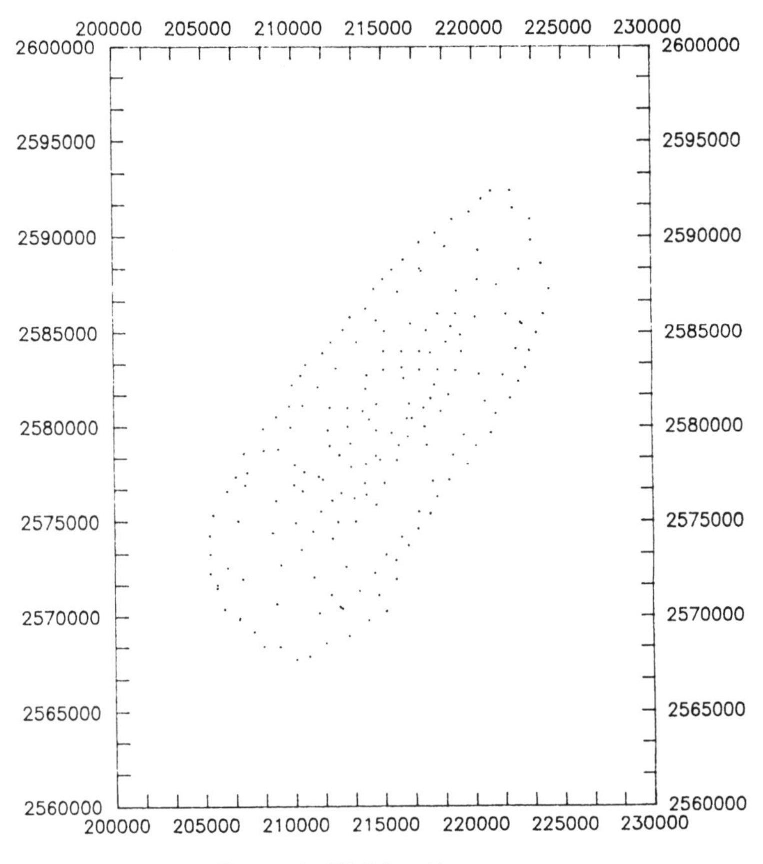

Figure 3. Well location map

(Greenacre,1984), the first subset is denoted "principal", controlling the reservoir quality zones, and the second, denoted "supplementary", provides the basis for a validation criterion construction. The degree of homogeneity of the supplementary variables within each zone is expected to be maximum.

Several contingency tables, which cross-tabulate the two principal variables for different class limits, were used as input into the correspondence analysis program. The results of this procedure for one input table are given in Figure 4.

Based on sample projections onto Axis 1, the splitting procedure was carried out on the output provided by applying correspondence analysis to each of the contingency tables to be tested. The resulting groups were compared from the standpoint of well contiguity and of the homogeneity of the supplementary variables. For the purpose of boundaries estimation, samples belonging to each group were coded as indicator variables, and the geometric variogram, reflecting the morphological structure of groups, was used as a basis for the transitive kriging procedure (Matheron, 1971).

To convert the indicator kriged values into a binary map, the same methodology of the previous example was applied. The estimated shape is given by the set of 1's of the binary map.

This iterative process produced the final zonation displayed in Figure 5 for the three main zones of the reservoir.

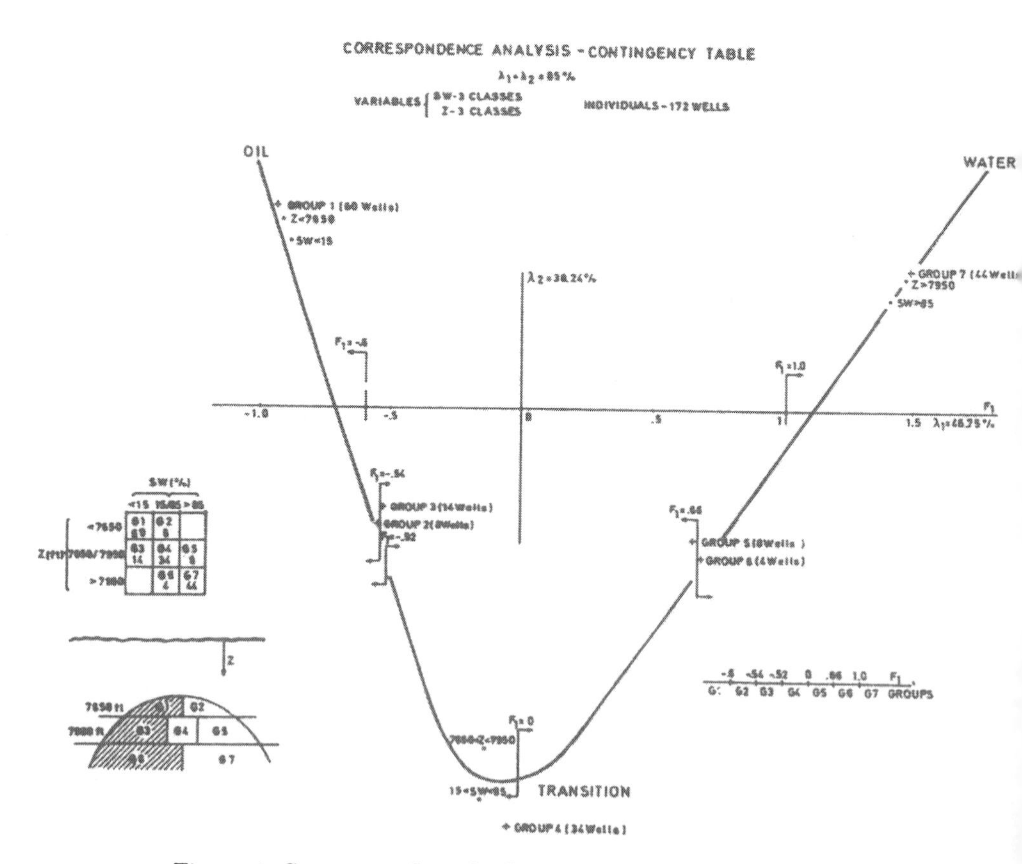

Figure 4. Summary of results from correspondence analysis

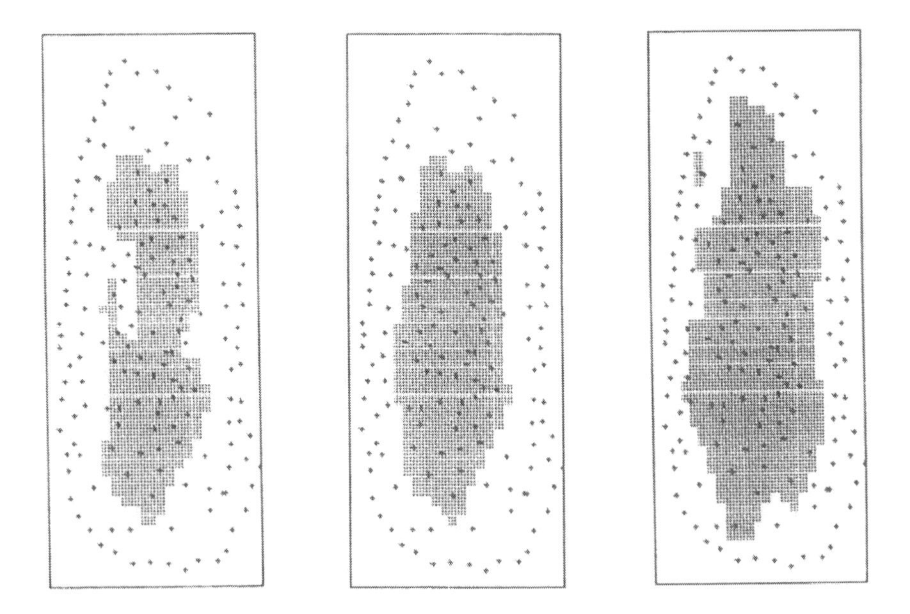

Figure 5. Results of zonation procedure for zone 1, zone 1+2, and zone 1+2+3

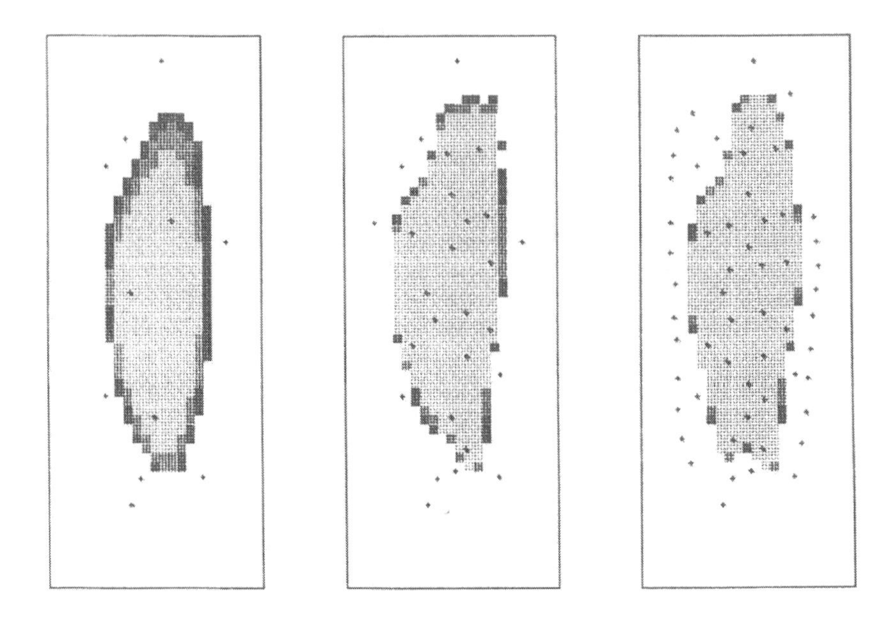

Figure 6. Boundaries estimation

In order to assess the influence of the information level on the performance of the method, a test was carried out for zone 1 using only 11, 27, or 56 wells. These correspond to three stages in the reservoir recognition process.

The results of this test are given in Figure 6, where confidence bands of the geometry produced by the estimated method also are displayed.

CONCLUSIONS

The zonation problem has been solved using a new methodology combining factorial analysis and geostatistics. The results of applying this to two case studies are given to illustrate its potential as a tool aiming at the interactive zoneography of mineral resources. In the sulphide orebody, three ore types which feed different metallurgical plants were recognized and mapped. In an oil reservoir, several quality zones supporting specific production strategies were found and their boundaries estimated and plotted.

REFERENCES

Greenacre, M., 1984, Theory and applications of correspondence analysis: Academic Press, London, 364 p.

Lebart, L., Morineau, A., and Warwick, K., 1984, Multivariate descriptive statistical analysis: John Wiley & Sons, New York, 510p.

Matheron, G., 1971, The theory of regionalized variables and its applications: Les Cahiers du C.G.M.M. de Fontainebleau, fasc.5, 112p.

Soares, A., 1990, Geostatistical estimation of orebody geometry: morphological kriging: submitted to Jour. Math. Geology.

TEACHING AND TESTING STRATEGY IN MINERAL EXPLORATION BY SIMULATION TECHNIQUES ON PERSONAL COMPUTERS

M. Vannier
Ecole des Mines de Paris, Fontainebleau, France

and

R. Woodtli
Geodidax, Lausanne, Suisse

ABSTRACT

The coming of a generation of personal computers with low-cost graphic cards and math processors, gives a new impulse to geological modeling. It now is possible to simulate rapidly and with accuracy complex geological environments with reduced hardware. Such models can be used for teaching mineral exploration or for simulating geological data that can be output easily and used in teaching related techniques like geostatistics or multivariable analysis.

Aside from teaching, several uses of complex geological models are promising such as the modeling of real targets in order to test various strategies of exploration or exploitation. Another use is the experimentation of natural processes such as erosion, dissemination of elements around ore-bodies, folding, faulting, etc...

Use of Microcomputers in Geology, Edited by D.F. Merriam
and H. Kürzl, Plenum Press, New York, 1992

INTRODUCTION

Our experiments in simulation started in 1965 and developed by degrees through the years, mainly as a teaching instrument. Today, they also are an attractive method of experimenting or modeling processes, either natural or artificial, in a given geological environment.

The main goal is the teaching of mineral exploration. Learners involved are students in geology and in mining engineering (last year) and professionals attending refreshment courses. The aim is to guide learners in understanding the diversity, the relative importance, and the interdependence of the numerous factors which play a role in decision-making when looking for minerals. They are led to apply their scientific knowledge to practical goals while taking into account technical and economical constraints. They have to interpret, correlate, represent, and use the large amount of information available during mineral exploration.

A SHORT HISTORIC

We started teaching by using the classical method of Case Studies, but we dropped it early for more active involvement of the students.

Relying on detailed data about a real gold ore deposit in Africa, we then supplied learners with real base material taken from a huge number of topographic and geological maps, logsheets, cross sections, chemical analysis, and so on. The method worked rather well but was limited to information actually available and thus was not satisfactory.

After having experienced these limitations, it became clear that we could give learners an opportunity to conduct their own exploration with the same constraints as in real life by using artificial data and simulating an investigation of an idealized ore deposit. Our previous attempts had shown that it was convenient to supply learners with geological information through logsheets and so this medium was selected as the main exploration method. We then had to prepare beforehand logsheets of simulated boreholes drilled in an idealized orebody. We began to introduce economical constraints by simulating the availability of drilling machines from where learners could rent material and take into account expenditures like the cost of drilling per meter. At the same time, we introduced technical constraints on drills such as maximum depths and speeds. The teaching period was divided into rounds of sessions, each session representing one month of field and laboratory work in real life but squeezed into a few days of simulated exploration. Although rough, the model was reliable and adapted to simulation of many aspects of real situations. The main obstacle was the time consuming task of preparing logsheets which thus prevented any sophistication of the geological model.

The next attempt was based on an analogical model for the geology. A geological model made of transparent sheets of plastic was used to simulate discontinuities between lithological formations. A drill, whose position could be adjusted precisely along the x and y axes as well as in direction and pitch, was used to make tiny holes through the structure. All intersections were reported on a logsheet. The main advantage of such an analogical model lies in its ability to be presented to learners at the end of the sessions, therefore giving them an opportunity to compare their abstract view of a three-dimensional block of data with a realistic representation of the geological space. Here again, some difficulties were present, the main one being the long task of performing a drilling.

In 1970, mainframe computers where readily available and seemed to be the obvious solution to our quest for accurate and quick answers. After some research, it became possible to simulate on a computer a sophisticated model of geological space as well as models for technical and economical constraints (Vannier and Woodtli, 1979). Logsheets were calculated and typed rapidly, allowing us to improve dramatically the sophistication of models responses. In fact, a new aspect of simulation clearly available at that time: the need of sorting the huge amount of data, a computer can output and the imperative that simulation must present an already pre-digested set of data. For instance the computer could output, instead of the name of a rock sample, its color, its hardness, its structure or even its chemical analysis and let the learner determine its nature, as is the situation in real life. Such an influx of data would have burdened learners without benefiting our goal which was to emphasize exploration and not petrology. Roughly speaking, one could say that it was possible to shrink one month of real work on the field into a few days of simulation with the only condition that many aspects of the real work has to be done by the computer itself.

For the first time it became possible to investigate mineralization more thoroughly and perform geostatistical evaluations of ores (Maignan and Woodtli, 1981).

The progress, compared to other attempts was impressive, but there were some drawbacks: mainly the chore of building such realistic models. An example may help to understand the complexity of a computerized model. One of our last models described an area of 30 km by 20 km wide and more than 2000m high. More than twenty lithological units were introduced with more than ten occurrences of complex mineralizations.

Faulting and several types of folding occurred in the prospect. A topographical surface had to be generated taking into account the superficial geology. In such a model, information concerning features as small as one meter wide could be retrieved and it was not conceivable to build it entirely by hand. The computer had to take some initiative in creating thousands of geological details. A primary model indeed was carved by hand. It contained primary tectonics and the computer was asked to add other phases of foldings, to introduce mineralizations in their respective environment, to create a realistic topography etc...

Because of this initiative given to the computer, it was difficult for designers to be aware of all the real aspects of the model and they had to drill the model everywhere in order to make sure that no geological inconsistency had taken place. When any inconsistency was discovered, the whole process had to be done again after a correction had been made inside the databank. In fact, some inconsistencies were discovered years after our model had been in use, fortunately without any serious hindrance for teaching.

The coming of the new generation of personal computers with their astounding capabilities in speed, memory, and cheap graphic boards has deeply modified our conceptions.

THE NEW GENERATION OF MODELS

The impact of personal computers has been twofold: a better control on what really is inside the model at the conception stage and, for teaching, a new approach by learners who now are able to process their data directly on their own computer (Jaboyedoff and others, 1989).

Depending on which type of participants are involved in the game, two different ways have been in use. The teaching period may cover one semester (16 weeks) with a one-half-day session a week and homework, or on the contrary, it may be concentrated into 3-4 weeks, with one or two rounds a day. In the first situation, lectures and seminars are included in the workshop, whereas in the second situation there is little time available for theoretical teaching and general discussions.

In both situations, learners are divided into teams of three people each, with the effect of introducing a feeling of solidarity within the teams and of motivating competition between teams. Each team has a small personal computer and communicates with the leader by the channel of floppy disks. Requests are written on the floppy, then processed on the leader's computer and answers are written back on the same floppy. The answers data can be alphanumeric or graphic and can be output on a local printer. When a computer for each team is not affordable, a simpler solution is to input requests and output all documents on the leader's computer. For years in the past, this was the solution we adopted.

At the beginning of a teaching period, learners receive the necessary initial information about rules of the game and some documents concerned with the geology and metallogeny of the investigated area. Their first task is to analyze and then to synthesize all information in order to select exploration targets and elaborate a strategy they will have to use given the time and resources allowed. Several games may be staged, depending on the initial knowledge of the prospect: either identification of mineralized zones, or selection of ore showings for further investigations, or systematic exploration including evaluation of reserves. Consecutive games may be linked together and learners in this situation have to refer to previous reports of predecessors.

Although drilling is the most important method of investigation, new tools have been included, especially geochemistry and recently the oldest geological tool ever used, hammer geology. Participants may order geological maps with indications on lithology and dips or geochemistry maps (stream sediments or tactic) for several elements. Geophysics prospection will be available in a near future, mainly as down-the-hole geophysics.

Usually, learners are free to select any method of investigation they want: tools, locations, depths of boreholes all are open options. They are free to use any procedure for processing their data: by using the computer or by more conventional ways such as maps or cross sections. The main interest in this type of teaching is to bring learners to discover by themselves, with the constant help of the leader, the best techniques adapted to their needs. For instance, our experiences has shown us that many students are overloaded at first with the huge amount of borehole data to handle properly until they are advised to use structural maps as an alternative to logsheets or cross sections. They also discover the interest of anticipating results when planning drillings; they are not allowed to change orders after their request is made, therefore requesting an adequate depth or location of a drilling may save money and time for other purposes. In addition, they discover that even some elementary theoretical views on mining geology may help to understand the relations between ores and regional or local geology and are pushed to make some investigations into the literature or to bring forth discussions with the leader. This simulation can be best seen as merely a method of impelling learners to ask questions and of putting them in a situation where theoretical, technical, and financial aspects of mineral research become acute topics and need to be investigated. These remarks point out the important role of the leader which is really an essential part of the game because he has to transform what could be a type of arcade game into a real simulation. Thus the role of leader can be played only by well-experienced people, having good knowledge of mining exploration or geostatistics, etc... depending on the goals.

In the situation of mineral exploration, simulated laws imposing a rhythm in the taking of claims force learners to make decisions. They have to give up some less promising claims and concentrate step by step on one or two locations.

At the end of the teaching session, the leader, colleagues, and guests from the mineral industries or even from banking are gathered in order to listen to and to appraise the presentation by each team of a report on their exploration. This report has to be ready a few days in advance so as to be read beforehand by the leader. It contains information on geology, the strategy selected and results and recommendations for any future work on selected prospects. This type of brainstorming session incites numerous suggestions and ideas and usually is as useful for learners as it is for the leader. It has been the source of many improvements for the games.

THE BUILDING OF A GEOLOGICAL MODEL

When confronted with the problem of modeling a geological space, the first thought was to build by hand or by simulation a databank where any information concerning an elementary cell could be retrieved rapidly, the data being accessed by the three coordinates of the cell. In a prospect of 30x30 km where gold mineralizations of 20 cm high can be intersected at a depth of more than 100 m, it would be necessary to store more than 1.0E+13 cells, each of them containing parameters about rock formation, grades of several elements, etc... It was clearly a wrong solution for many reasons, even if this huge theoretical number of elementary cells could have been considerably reduced by using more sophisticated techniques.

Models used in teaching and models used in simulating processes are not always compatible. Precise modeling of processes, such as erosion, needs long calculations and although they give realistic results, they can not be used always to build a geological model for teaching exploration. For teaching purposes, models have to be quick and a compromise between speed and realism has to be determined.

Geological models used to simulate a three-dimensional space now will be discussed through the example of Claim. The main part of Claim is a function written in C language whose name is “fore”.

Fore is in itself a simulated geological model where any vertical drilling can be performed in an area of 30x30 square kilometers and for an unlimited depth. The topographical level is calculated by this function. More than five types of drillings are allowed: percussing, core drillings, dry drillings, augers, and special drillings, each of them leading to differents answers. For instance, percussion drills usually are faster than core drills but give less precise information on geological levels or grades and do not show distinctions between certain types of rocks. Special drillings are used to get information on the last eroded portion in connection with the simulation of geochemistry or other geological data.

This function uses a databank where all information concerning lithology, mineralization, topography, weathering modules of rocks, etc... are stored. As the databank can be changed easily by qualified people, it is possible to use fore with different databanks and produce different geological environments. This function has been devised to work rapidly even on a personal computer and is used for a lot of purposes that may need thousands of drillings such as the mapping of geology.

The first task of fore is to calculate every intersection of a given vertical with surfaces representing discontinuities. A discontinuity can be a limit between two lithological units or it can be a fault or a facies change. Even extrapolated intersections have to be calculated in order to fill the whole space with accuracy. Mathematically, these surfaces are described in several ways depending on the nature of the discontinuity. A surface can be described by a succession of cross sections, by triangulation, or by mathematical formulae. It may require two or more steps to generate a surface; a first step for the general shape and subsequent steps for small-scale details. For instance, the description of an old eroded basement can be produced first by cross sections and then by adding mathematical distortions to simulate hundreds of rolling hills that occur in such a paleotopography.

The second task of fore is to eliminate all virtual parts such as extrapolated portions of surface through faults for example. This method of calculating intersections between geological discontinuities is powerful and signified a breakthrough in our geological simulation when first derived.

At that point of the calculation, it is possible to introduce several phases of faulting and folding each of them following a style that can be imposed by the computer.

The next stage is to calculate ore grades and paragenesis, if any. The simulation of ore grades is made by piling three-dimensional "clouds". A cloud can be considered as an ellipsoid defined by four parameters in addition to its center coordinates: its length, width, height, and grade present at its center location. A mathematical law is given to describe the way grades are fading away when the distance from center is increasing. It can be positioned either directly or inside a defined geological unit prior to any tectonic event. Other mathematical shapes have been used to simulate ore grades such as toroidal ones, depending on the type of mineralization (Lanoe, 1981). When this first stage of calculation has been done, a second stage consists in adding geostatistical perturbations in order to simulate small-scale details within ore-grade distribution. Usually a predefined variogram is superposed to the first stage distribution. Paragenesis are simulated in a similar fashion.

The final task then is to create a landscape in accordance with the geological underground (Fig. 1). We determined that the easiest way to perform such a complex task was to use a simplified simulated erosion process taking into account a mechanical characteristic for outcropping rocks. It would be possible to simulate a realistic erosion process but, as we said earlier, the model has to answer quickly and cannot take into account data from distant areas without long calculations. The simplified erosion process is done in two steps: a first step removes all rocks above a predefined smooth topography and a second step removes rocks starting from this first soil and according to specific parameters associated with all geological formations. The first topographic level is calculated by interpolation of an arbitrary set of data simulating a gently slanted landscape and is quickly available. For the second step, a random regionalized distribution is added to the mechanical parameter in order to simulate variations of a rock's response to erosion. A special process of calculation is used to simulate small scale topographic details in undifferentiated geological formations as, for instance, basement complexes that do not need to be known precisely in certain applications. In that situation, a type of fractal generation is selected, but for efficiency reasons, a noniterative mode of calculation has been derived that uses a Fourier's development.

Using a similar procedure, several alteration levels are created starting from the topographic level. A bedrock is first calculated leading to what is termed a soil layer. When this layer is thin, it is alleged that an outcrop occurs at this location. A weathered zone then is created. This zone cannot be sampled by usual drillings. A third zone termed washed zone, is where sulfide mineralizations are subject to oxidation.

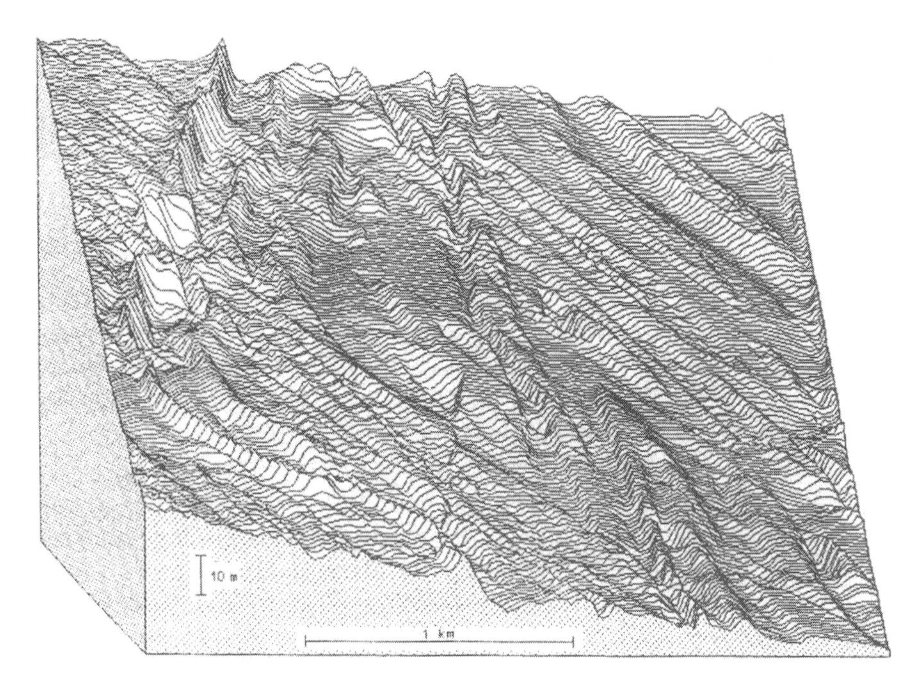

Figure 1. Simulated landscape. Altitudes are highly exaggerated. This area shows contact between metamorphic complex (left upper corner) and overlaying folded sedimentary series. Fault is visible. This type of picture is not available for learners.

In addition to the basic function fore, a number of programs have been written. Some of them use the fore function, for instance the geochemical model or the geological map model and some do not, for example the logistic model which is handling budgets, expenditures, and tools.

MODELS USED IN A GAME

The first use of the fore function is a program to simulate drillings, either vertical or inclined. A first type of drillings uses augers to perform vertical boreholes not deeper than the bedrock and that can be dispatched on arrays. Other types are percussion drillings and core drillings for information below the bedrock and dry drillings for sampling into the weathered part. When mineralizations are encountered, chemical analyses are systematically performed.

The geochemical model simulates exploration using geochemical sampling. Sampling is done on a regular array and takes into account backgrounds of all lithological units and grades of ores if any, either in the immediate underground or in the last eroded part (Fig. 2). Topography is used to simulate a diffusion of metals down the slope. Data can be output in different ways: through listings or maps with or without contours printed on conventional dot-matrix printers. All geochemical backgrounds can be modified easily by the leader. Stream sediment geochemistry is done by hand, taking into account data from the geochemical model and topographical data concerning the hydrography.

The geological map model creates geological maps at a given scale, with information on lithological units and on dips (Fig. 3). It is possible to create maps restricted to outcrops such as those simulated by the weathering process or special maps stripped of the soil layer. Outputs are sent to a dot matrix printer. All maps use a graphic key for each geological unit, avoiding the need of color printers and copiers. Three types of

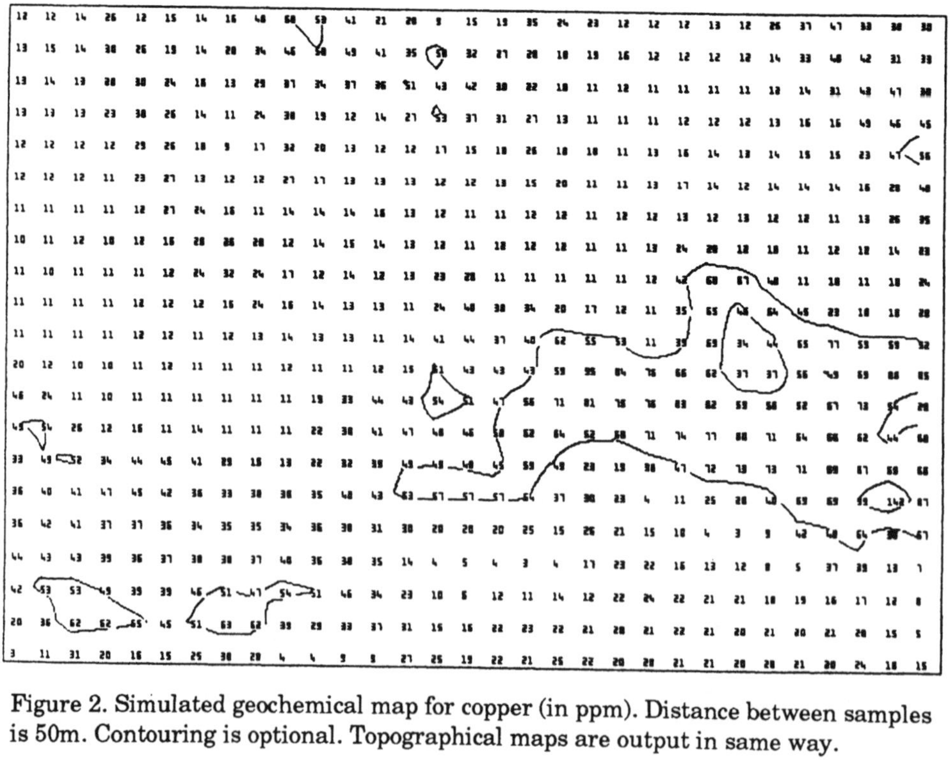

Figure 2. Simulated geochemical map for copper (in ppm). Distance between samples is 50m. Contouring is optional. Topographical maps are output in same way.

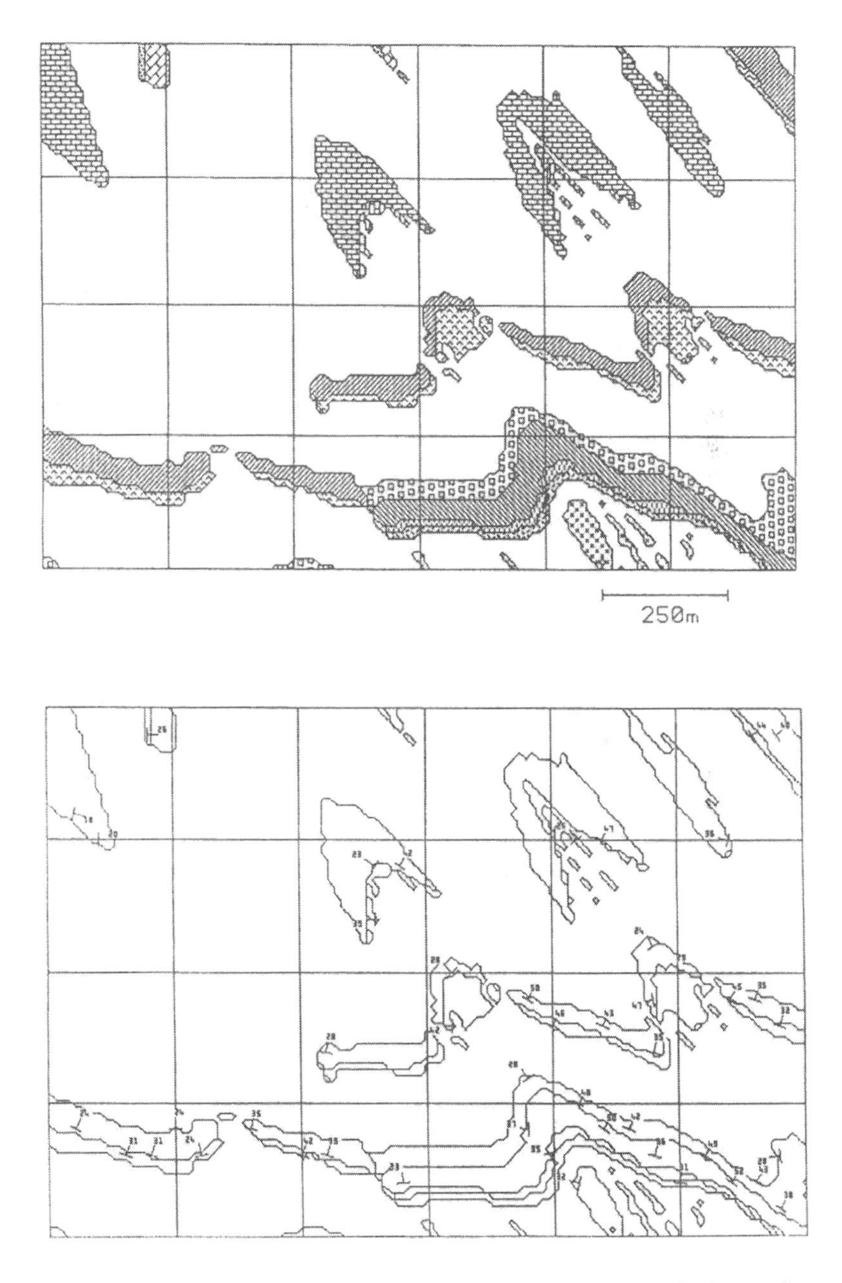

Figure 3. Lithological map and dip map restricted to outcrops. Scales and type of mapping are optional.

mapping are allowed: mapping of all geological formations, mapping of main geological units, and lithological mapping. Usually, teams receive only dip maps and lithological maps of outcrops, the other types of maps being reserved to the leader. Graphic keys can be modified easily, each design being selected among more than one hundred possibilities.

The logistic model manages all tasks concerning teams' budgets, leasing of drilling machines, and boreholes' location archives. It keeps a statistical archive of each team's way of spending its financial resources. It is possible to know how much money has been spent on leasing material, on boring with any given type of drill, on geochemistry or geological maps and so on. Usually teams are allotted a first budget covering expenses for three or four periods. They have to provide sound reasons that justify their needs in order to receive more funds. It should be noted that this logistic model can be bypassed for games that do not need financial guidance.

TOOLS FOR MANAGING

The leader has access to several managing softwares through a menu. The first one used is the initialization of a game session. Prior to this initialization it is possible to indicate the number of teams and to change some default data about drilling machines allowed, costs of leasing, drilling, moving, etc...

Other commands concern checking floppy requests prior to their processing, processing requests, and allotting budgets.

The leader has restricted access to a number of tools which are devised to help him understand rapidly any problem arising during the game. It can be a pure geological problem such as understanding a complex structure or checking any strange interpretation of data by a team. It also can be a financial problem, such as bad management of resources by a team which is running out of money.

For geological investigations, a fast program giving random cross sections at any given scale with or without information on ore locations and grades is provided (Fig. 4). This graphic software needs a conventional graphic screen (CGA 320x200 type) and can be used easily to retrieve any information. Another tool is a software able to produce small scale geological maps not usually available to learners. Results can be output on a graphic screen (VGA 640x480 type) but also on a dot-matrix printer.

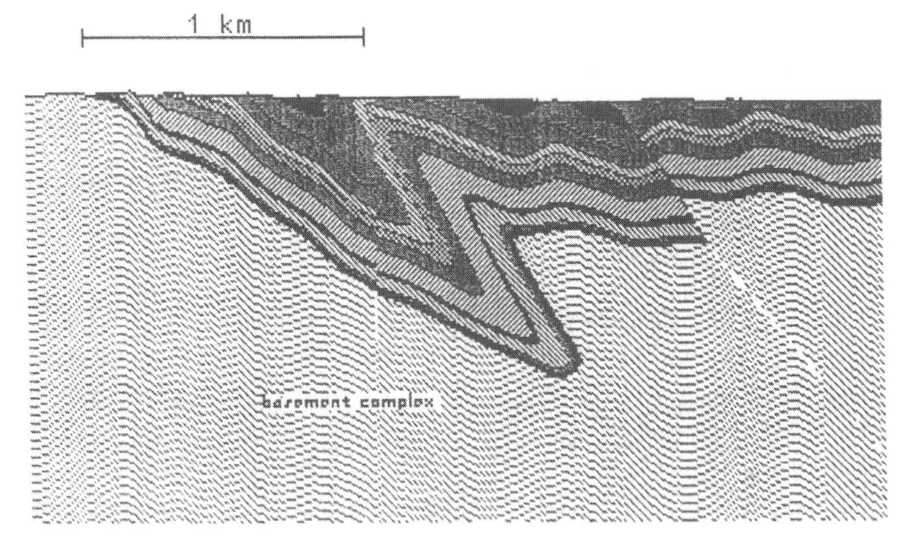

Figure 4. Cross section, as it appears on low-resolution screen. It is possible to edit drawing and to display rock names before printing. Mineralizations can be indicated when superior to requested grade.

For logistical control the computer can output any information needed from the past periods. All team requests are kept on the leader's hard disk and can be consulted any time as well as statistical data on financial resources.

LEARNERS TOOLS AND EQUIPMENT

When the solution of providing each team with a computer is selected, several basic softwares are accessible through a menu on each local computer. One concerns the filling of requests and can be viewed as an editor. Its main function is to perform a first level control of requests syntax. Other commands concern the handling of answers and perform simple functions like the printing of alphanumeric or graphic data (maps).

Depending on the aims, other softwares may be installed similar to conventional statistics or more sophisticated tools such as geostatistics for ore evaluation. It may be useful to install a text processing software and a small documentary databank where some literature concerning the geology and economy of related ore deposits can be accessed.

Each team disposes of a private room where documents can be laid out easily and safely and where discussions can be carried freely on without interference from other teams. It is important to preserve a spirit of competition between teams and to encourage discussions and criticism inside teams or between teams and the leader. As in real life, the composition of a team may have tremendous effects on its ability to handle the task and it is studied carefully beforehand.

COMPUTER REQUIREMENTS

Minimal computer hardware requirements are small. For the leader, a simple PC working under DOS (or XENIX) operating system with a hard disk and with 520 ko (720 ko) of RAM is sufficient. In practice, it is necessary to use an arithmetic coprocessor in order to minimize computing time. A dot-matrix printer also is needed to produce documents used during discussions with learners (financial documents, for instance) or as working tools. A graphic screen of fair resolution similar to VGA greatly helps to understand any geological problem in a prospect as mentioned earlier. Software is for the 80286 processor but also work on the 80386 processor.

For teams, equipment depends mainly on the type of data processing needed by learners. For example, if learners are required to use geostatistical methods for ore evaluation, an arithmetic coprocessor is needed. A graphic screen of moderate resolution, such as CGA, also may be useful. Softwares are built for DOS operating system and for 8086 and 80286 processors. Although less convenient, a solution without computers for the teams is possible, all requests and documents being prepared on the leader's computer.

The game also could work on a mainframe computer running under the UNIX operating system. Teams use terminals instead of personal computers.

CONCLUSIONS

A model of a geological space can be used for teaching various topics such as mineral exploration, evaluation, and exploitation of ore deposits. It can be used also to simulate complex processes such as erosion or geochemical diffusion of elements. Depending on the goal, different ways of building a model are possible, the main considerations being the speed of calculation and the sophistication of the hardware. These are crucial elements in teaching experiments but good compromises between realism and efficiency are possible. Personal computers offer speed and now, fair graphic capabilities at low costs. They are suited perfectly for such tasks.

ACKNOWLEDGMENTS

We are indebted to P. Drucker for many improvements of our manuscript.

REFERENCES

Jaboyedoff, M., Savary, J., Bauchau, C., Maignan,M., 1989, Teaching mineral prospecting by personal computer-assisted simulation: Hamza, M.H., ed., Proc. IASTED Intern. Symp. on Modeling, Identification and Control (Grindelwald, Switzerland), p. 47-50.

Lanoe, S., 1981, Modèlisation sur ordinateur d'un compartiment géologique et applications pédagogiques. Exemple des gisements de cuivre porphyriques: Thèse Ingénieur Docteur, Ecole des Mines de Paris, 150 p.

Maignan, M., and Woodtli, R., 1981, Geostatistical analysis of computer simulated ore deposits: Proc. First Intern. Conf. on Modeling and Simulation (Lyon, France), v. 4, Materials and resources, p 77-84.

Vannier, M., Woodtli, R., 1979, Teaching mineral prospecting by computer-assisted simulation techniques: Computers & Geosciences, v. 5, no. 3/4, p. 369-374.

COMPUTER-AIDED DECISION TECHNIQUES FOR HYDROGEOCHEMICAL URANIUM EXPLORATION

Patrice Poyet
Centre Scientifique et Technique Du Batiment, Valbonne, France

ABSTRACT

Multivariate analysis has been recognized as a powerful tool in geochemical exploration, but special emphasis also has been given to the necessity to complete adequate preliminary investigation of the data before attempting to use sophisticated data processing and manipulations. The methodology presented here relies on extensive experience with multivariate methods gained from the processing of large case studies coming from surveys carried out by French mining companies using hydrogeochemical uranium exploration. We have developed a robust methodological approach and a set of integrated software available on microcomputers to model the distribution of elements in water analysis and to account for the mixing of the geochemical end-members observed and to tackle the definition of an adjustable modeling of background compositions and of their related anomalies after the removal of the disturbing outliers from the recognized statistical populations has been achieved. The policies used lead to an efficient set of data processing, data integration, and data representation software making it possible to offer practical assistance to the exploration geochemist when faced with decision-making processes.

Use of Microcomputers in Geology, Edited by D.F. Merriam
and H. Kürzl, Plenum Press, New York, 1992

INTRODUCTION

Geoscience data have become huge sets since the progress of the modern analytical laboratory led to low cost comprehensive multielement information for geochemical surveys. This explosion of data gathering has been noticeable especially in the context of hydrogeochemical exploration where tens of variables can be measured for hundreds of samples covering the abundance of major or trace elements in water analyses. An immediate consequence is that this enormous volume of data precludes any relevant thorough processing without the computer assistance spanning from the data collection and static organization thanks to relational or object-based database systems, the processing using univariate, multivariate, or complex numerical analysis methods, the representation based on high resolution displays and printers, the integration of high level highly significant model results, to the assisted decision-making process involving sophisticated programs based on advanced software engineering techniques such as expert systems. The widespread use of powerful 32 word bit based microprocessor units rendered possible this extraordinary evolution in the geochemical information management and lead us to develop an integrated set of programs to cope with the management of hydrogeochemical data collected during the water geochemical surveys of the Corbigny and Lodeve (France) reconnaissance programs devoted to locate blind uranium deposits through indirect prospecting based on water samples collected in drillholes. The methodology developed to achieve the data modeling phase to account for the background composition and the related identification of relevant anomalies was validated using data collected in the Lodeve area where an important mineralization has been recognized for a long time and where important on-site mining work led to a comprehensive knowledge of the three-dimension geographic distribution of the orebody. These techniques were applied later on to the Corbigny area (Morvan, France) where it was possible to evaluate their significance as a prospecting scheme, and confirmed the discovery of a mineralization of limited extent. As a brief introduction to the method developed, let us present the main treatment steps. First, a data segmentation phase leads to homogeneous units, for which a mixing model is proposed to compute the geochemical background and to appraise for the different types of groundwaters involved in the related proportions of the end-members. According to this iterative method (many iterative steps can be achieved to obtain correct subsets) it is possible to explain some part of the observed concentration for the trace elements thanks to the previous mixing model based on major constituents and then to assess second-order anomalies which can be identified as representing the unexplained part of the measured concentrations. Applying this framework with careful and knowledgeable policies (concerning the transformation of geoscience data responses to geochemical processes) seemed to be an efficient prospecting scheme for the discovery of blind orebodies using indirect prospecting methods such as groundwater geochemistry. First, we describe the methodology proposed to recognize significant multivariate anomalies, then we illustrate the discussion throughout with the processing of a set of real data drawn from the Corbigny area. A brief introduction to the geological context is provided, main results are reported and a description of the original use of multivariate methods is made when necessary.

DATA-PROCESSING FRAMEWORK

Basic Layer

When the exploration geochemist faces a survey, the first task is to gather the data and to organize them in a collection of items according to a formalism well suited to ensure their further retrieval. Much software has been developed to cope with these requirements such as the GRASP program of the USGS or the FFG system we designed a few years ago, generally making it possible to select samples according to predicate calculus based on logical or arithmetical requests. But we must observe that computer scientists have tackled this problem for a long time leading to the design of extremely efficient fault tolerant environments able to cope with huge amount of data, ensuring moreover data integrity and coherency and defining normalized formalisms to achieve data requests based on algebraic ensemblist languages such as the SQL standard for relational models. Recent developments in the domain of object-oriented databases are really promising and aim to merge deductive capabilities coming from the artificial intelligence field and massive data-storage facilities issued from the database technologies. This is one of the main reasons why we will assume that some database systems are used by the geochemist without describing this layer further (our files are processed by the FFG) as current technologies are becoming obsolete quickly. Let us consider that this layer provides for usual database functions such as storage, retrieval capabilities based on operators such as selections, projections or joints for relational environments, update and of course usual statistical functions most of the time including comprehensive packages to achieve at least univariate analysis. Once basic data have been collected and digitized into a database computer system, it is a wise strategy at an early stage of data treatment to carry out some form of analysis to determine the morphology of the data. It now is well known that if the original data can be subdivided into one or more homogeneous statistical subsets, it would be far more meaningful to carry out a series of separate analysis after having discriminated the subsets into sets of clear geological significance (Sinding-Larsen, 1975; Tuckey, 1979).

Data Segmentation

According to technical and geochemical considerations, it is difficult and indeed not relevant to process as a bulk of information trace elements and main components as their respective concentration ranges are different. Geochemical laws accounting for the observed concentrations of trace elements (Rn, He, U) differ from those involved to explain the abundance of main components such as SiO_2, Ca^{2+}, Mg^{2+}, Na^{+}, K^{+}, Cl^{-}, $SO4^{2-}$, HCO_3^{-}, or F^{-} (referred later on as major elements) measured in water samples collected at drillholes. This observation leads us to model the concentrations of the major elements first, and then to provide an interpretation scheme based on the results of the former model to explain trace values. Moreover it should be stated that whereas results of interpretation of an entire data set may indicate vague generalized relationships or broad element associations, if such results have been obtained using heterogeneous data populations including outliers, the observations are not relevant and can be invalidated at the detailed level (Howarth and Sinding-Larsen, 1983). Thus, the first step, of course, is to remove obvious outliers from the data set and to study them apart. The next stage is to identify homogeneous composition sets thanks to a clustering method and to characterize them using basic statistical parameters such as means, standard deviations, and analyzing the sample corpus and especially remote samples from the center of the classes. Many methods can be used to achieve clustering (Duda and Hart, 1973; Everitt, 1974) and a review is presented in Cormack (1971) and Tryon and Bailey (1970), but we select the "dynamic cloud method" developed by (Diday, 1975) for the sake of robustness and as it ensures an easy interpretation of the results. This typological method assumes that we know, a priori, how many subsets the population will have to be subdivided into and to pick at random a sample to form the center of the first group. The center of the second class will be the sample lying the farthest away from the first one (according to any given distance criterion) and so on, selecting the remaining centers at maximum distances away from those already allocated. Samples then are assigned to the group to whose center they are the nearest. If x_{ik} stands for a set of measurements on the ith sample in group k, and there are n_k samples assigned to that group, the group average or its center of gravity is given by (1):

$$\bar{x}_k = \frac{1}{n_k} \sum_{i=1}^{n_k} x_{ik} \tag{1}$$

For each composition set, it is easy to compute a measurement of the scattering of the data within this group (2):

$$W = \sum_{j=1}^{k} \sum_{i=1}^{n_k} (x_{ik} - \bar{x}_{ik})^2 \tag{2}$$

As long as the process iterates, samples are attached to the groups so as to minimize the criterion W. The gravity center of each group is then recomputed, and substituted to the old one. Group centers move until a stable position minimizing W can be reached and lead to recompute for each iteration the distances of the samples to the new group nucleii, a complete process representing "a trial". Many successive trials are achieved and the results compared until robust shapes can be obtained for the groups, thus leading to stable collections of samples. Let us illustrate this approach with the real sampling available for the Corbigny area. We suspected that eight distinct groups of water compositions could occur and the results for the set of 58 samples appear as follows (Table 1):

Table 1. This table shows partitioning obtained using "dynamic cloud method" for eight classes of water compositions for major elements measured

Var	Pop	Class1	Class2	Class3	Class4	Class5	Class6	Class7
He	70.1	6.6	4.5	6.9	9.1	112.8	268.4	325.7
F^-	2.0	1.2	0.5	0.7	2.6	2.6	3.5	3.5
HCO_3^-	270.0	400.1	353.2	405.9	193.1	214.5	66.9	224.6
$SO4^{2-}$	213.6	1196.3	41.5	529.2	39.7	63.4	250.0	61.6
Cl^-	66.0	51.2	18.0	41.5	25.8	158.7	361.7	334.4
SiO_2	12.8	9.9	10.9	12.0	14.3	13.3	8.3	14.6
Ca^{2+}	128.7	455.9	106.1	234.1	56.9	60.9	216.5	71.0
Mg^{2+}	22.0	72.3	13.0	60.1	10.5	12.4	0.2	12.7
Na^+	61.8	40.6	13.0	46.3	18.3	116.6	42.5	246.4
K^+	9.6	24.2	7.2	11.0	6.3	8.1	37.2	9.3

Without providing detailed explanations of these results which will be given further when considering the Corbigny case study, one should notice three particular composition poles: the extreme sulfated composition represented by class 1 including two remaining outliers (refer to Table 2), a chloridized pole modeled by class 7 (we also could assign to that group the sample individualized in class 6), a bicarbonated calciferous pole represented by class 2 (and also class 4), and a geochemical trend characterized by a composition enriched in chloride and sodium spanning from class 5 to class 7. The immediate issue of course is to consider the geographical distribution of the observed classes and the various maps produced are discussed in the related section. Interesting information also can be deduced from the analysis of the position of each sample within its associated class given its distance to the class center, and from various representations characterizing the compositions of the classes. This first appraisal of the data structure is relevant to assess the exact diversity of the compositions encountered which should be related to geochemical processes (here mixing models), and also can be used to split the data population into various homogeneous subsets reused for further modeling. Of course, such preliminary results should be confirmed thanks to a cross checking using complementary methods. Q mode factor analysis seems to be an extremely powerful tool, both to confirm the previously recognized end-members (i.e. contributing to a relevant data partitioning), but moreover offering a framework to account for mixing models, of special interest in our situation where each sample is a mixture of various water end-members coming from different aquifers. Once extreme compositions have been discarded, outliers removed and studied apart, it is possible to model mixing problems and especially to compute the contribution of each aquifer to the global composition of the sampled water. The process can iterate until homogeneous subsets can be obtained, either to be able to explain the trace compositions in terms of the mixing model elaborated for the major constituents (thanks to multiple regression analysis) or to determine adjustable trace thresholds according to the composition stratum characterized by major elements.

Mixing Models

Factor analysis has become widely used by geochemists, but this extremely powerful multivariate method leads in many situations to poor results because of misunderstanding of the underlying assumptions coupled to the processing of heterogeneous sets. Nevertheless interesting results were obtained using these techniques to solve mixing problems such as petrologic ones, and we deemed the outcomes so promising and our problem close enough to mixing formulations to use Q-mode factor analysis extensively and to provide a new interpretation framework to the technique in the context of water mixing problems. Factor analysis encompasses a broad panel of techniques ranging from Principal Components Analysis (PCA), Q-mode and R-mode Factor Analysis (FA), True Factor Analysis (TFA), and among other methods Correspondence Analysis (ACP). References given here cover most of these topics, and we now will focus the presentation on Q-mode factor analysis, especially well suited to the modeling of mixing problems such as those arising in water-mixing issues. The aim of factor analysis, whatever the underlying mathematical model may be, is to assess relationships occurring in a set of variables or samples and to summarize the scattered information in a set of factors representing new variables able to express clearly interrelations happening in the data set. Factors are theoretical variables, obtained as a linear combination of original variables, and computed so as to account for a significant part of the observed variance or covariance. Variables and samples can be considered as vectors in a N-dimensional space, where the data matrix is X(Nxp), N being the number of samples and p the number of variables. Factor analysis aims to reduce the dimension of the initial space, to summarize the valuable geochemical information. The starting point of these methods is to factor the data matrix, determining the rank of the matrix which is the smallest common order among all pairs of matrices whose product is the data matrix. It suggests that the data matrix can be regarded as the product of two matrices of which the orders can be significantly less than that of the original matrix, thus replacing the data in a smaller subspace whose dimension corresponds to the number of linearly independent vectors (i.e. the rank of the data matrix). This basic structure of the data matrix can be obtained thanks to the Eckart-Young theorem, stating that any arbitrary rectangular matrix can be represented as:

$$X = V \Gamma U' \tag{3}$$

- where X is the data matrix;
- V is the matrix of eigenvectors of the major product of order N x p;
- G is the diagonal matrix of singular values (i.e. square roots of the eigenvalues) of order p x p;
- U is the matrix of eigenvectors of the minor product (i.e. inner product) and U' its transpose, square orthonormal matrices of order p x p.

The analysis of XX' corresponds to Q-mode factor analysis which aims to account for interobject relations, and analyses the major product of the data matrix by its transpose. The aim of factor analysis is to approximate in a least-square sense a matrix W representing an approximation of lower rank of the initial data matrix X, such as:

$$W = V_k \Gamma_k U'_k \tag{4}$$

where: V_k is made of the k first columns of V, G_k is made of the k first rows and k first columns of G, and U_k is made of the k first columns of U, so that:

$$X \approx V_k \Gamma_k U'_k \tag{5}$$

Post multiplying Equation (5) with $U_k\, G^{-1}_k$ leads to:

$$V_k = X\, U_k\, U_k \tag{6}$$

where V_k stores the coordinates of the points in the new factorial space of reduced dimension k. Let us now give the formulation of the general factor analysis model, and then provide more detailed consideration covering the Q-mode calculus. Any factor analysis methods rely on the following mathematical expression:

$$X_{(Nxp)} = F_{(Nxk)} A_{(kxp)} + E_{(Nxp)} \quad (7)$$

and a solution to this formulation can be obtained factoring the data matrix X as the result of the product of the factor scores matrix F with the factor loadings A, making it possible to represent the samples in a subspace of dimension k, E accounting for the errors observed when reconstructing the samples in the reduced factorial space. Using the Eckart-Young decomposition given Equation (3) and neglecting the error term, Equation (7) can be rewritten as:

$$\hat{F}\,\hat{A} = V_k\,\Gamma_k\,U_k \quad (8)$$

and a possible solution is given by:

$$\begin{aligned} \hat{F} &= V_k \\ \hat{A} &= U_k\,\Gamma_k \end{aligned} \quad (9)$$

Q-mode factor analysis seems to be a well-suited technique when the objective is to classify (in some sense) a sample of observations (i.e. measurements in our situation) on the basis of several compositional properties. Q-mode factor analysis lies in the definition of interobjects (i.e. occurrences, observations or measurements, elements of the sampling) similarities, expressed thanks to an N x N similarity matrix containing the degree of similarity (according to any relevant assumption) between all possible pairs of N occurrences. The degree of similarity between two objects may be evaluated in relation to the proportions of their constituents, and for any two observations n and m of the data matrix (i.e. row vectors), the coefficient of proportional similarity (computing the cosine of the angle between the two row vectors as situation in p-dimensional space) usually is defined as:

$$\cos\theta_{nm} = \frac{\sum_{j=1}^{p} x_{nj}\,x_{mj}}{\sqrt{\sum_{j=1}^{p} x_{nj}^2 \sum_{j=1}^{p} x_{mj}^2}} \quad (10)$$

permitting to compute cos q for each possible pair of samples, and to arrange conveniently them in a N x N matrix of associations H, representing the matrix to be factored, the eigenvectors of this matrix providing a set of linearly independent reference vectors to which the samples (i.e. row vectors of X) may be referred. Row normalization, ensuring the removal of the effects of size differences between samples, can be achieved for the raw data matrix X using Equation (11):

$$W = D^{-\frac{1}{2}} X \tag{11}$$

where D = diag (XX'). The association matrix H, can be defined as the major product of the row-normalized data matrix, thus leading to Equation 12:

$$H = W W' \tag{12}$$

Observing that H is a real symmetric matrix, we have HU = UL, where L is a diagonal matrix of the eigenvalues and U is the matrix of eigenvectors ordered as columns, and U being square orthonormal so that U'U = U U' = I, this lead to Equation (13):

$$H = ULU' \tag{13}$$

According to the solution assumed by Equation (9), we can state that:

$$\begin{aligned} \hat{A}\,\hat{A} &= U_k\,\Gamma_k\,\Gamma_k\,U_k \\ \hat{A}\,\hat{A} &= U_k\,\Gamma_k^2\,U_k \\ \hat{A}\,\hat{A} &= U_k\,\Lambda_k\,U_k \end{aligned} \tag{13 bis}$$

which indicates according to Equation (13) and to Equation (14), that AA' = H, and leads to:

$$\hat{A}\,\hat{A} = \lambda_1 u_1 u_1 + \lambda_2 u_2 u_2 + \ldots + \lambda_k u_k u_k$$

In matrix form Equation (14) L_k is a diagonal matrix storing the k largest eigenvalues of H. Finally we can establish that according to the decomposition of Equation (3), the factor analysis model described by the

Equation (7), the simplified model of Equation (8) neglecting the error term, the hypothesis emitted Equation (9), the solution for the factor loading matrix can be deduced from the matrix decomposition of H (Equation (13)), and can be given by Equation (14):

$$\hat{A} = U_k \Lambda_k^{\frac{1}{2}} \tag{14}$$

which can be computed as the matrix of eigenvectors of H scaled by the squared roots of the eigenvalues of H. Moreover W, can be expressed according to the model of Equation (7), and omitting the error term as:

$$W \approx A F \tag{15}$$

where A is the N by k matrix of loadings and F is the p by k matrix if scores, and k is the approximate rank of W. Thus premultiplication by A' gives:

$$\hat{A} W \approx \hat{A} \hat{A} F \tag{16}$$

and observing that:

$$\hat{A} \hat{A} = \Gamma_k U_k U_k \Gamma_k = \Gamma_k^2 = \Lambda_k \tag{17}$$

we can deduce a solution for the matrix of the factor scores:

$$\begin{aligned} \hat{F} &= \Lambda^{-1} \hat{A} W \\ \hat{F} &= W \hat{A} \Lambda^{-1} \end{aligned} \tag{18}$$

One of the main objectives of Q-mode factor analysis then is to determine the number of independent end-members k, determining the rank of the normalized data matrix $W_{(Nxp)}$ or that of its major product moment $H_{(NxN)}$. Moreover it may be possible to get a close approximation of W with significantly less factors, thus representing true compositions mixing to account for the observed measurement according to the proportions expressed by the factor loadings. Q-mode factor analysis, therefore, can be viewed as an attempt to reduce a data matrix into a smaller matrix that may facilitate interpretation in terms of the mixing of compositional end-members, such as solution or precipitation of material that tends to be of a specific composition (i.e. pole). The data matrix is resolved into a concise model for which each analyzed sample is modeled as the mixture of a small number of factors reconstructing original material sources. Moreover, the extended form of factor analysis we use (Miesch, 1976), facilitates the interpretation of the results as they are expressed in the same units as the original data (otherwise they are dimensionless), and the composition loadings (refer to Eq. (14)) sum to unity and can be considered as mixing proportions.

Factor analysis also is a straightforward technique for classifying samples, and can be used immediately to achieve cross-checking of the results previously obtained with the clustering phase (Table 1), making it possible to reliably distribute original data in homogeneous subsets for which an accurate description of the compositional variation in sample suites can be achieved. For the same case study corresponding to Table 1 (i.e. the Corbigny area), we carried out a Q-mode factor analysis to give confirmation further to the water classes observed thanks to the clustering technique, attempting to explain with few factors a significant part of the variance of the data, and outlining relevant compositional end-members, well suited to evaluate the proportions of the mixing responsible for the samples values. According to the analysis of the factor-variance diagram based on the coefficients of determination, the variability in this data set can be accounted for reasonably by the three factor model of Table 2:

Table 2. Factor scores obtained thanks to Q-mode factor analysis for mixing model of Corbigny area (three more significant factor scores)

He	-7.180923	30.347321	0.183632
F-	0.868396	0.414385	0.476858
HCO3-	96.345291	6.362858	-4.847492
SO42-	-17.986694	5.954410	79.984940
Cl-	-0.741257	30.414734	0.109348
SiO2	6.308267	0.974594	-2.505931
Ca2+	19.128326	2.016218	21.822952
Mg2+	2.016218	0.694438	5.241902
Na+	-0.446445	21.873993	-0.266135
K+	1.688759	0.892729	0.753642

Given the factor scores representing the theoretical compositions involved in the mixing problem (Table 2), and the associated factor loadings accounting for the proportions of the factor scores for each sample of the data set, it is possible to reconstruct the composition of the j th component for the i th sample applying Equation (19), where a_{ik} represents the proportion of the k th factor for the i th sample (i.e. loadings), and f_{kj} models the abundance of the j th component for the k th factor (scores):

$$\hat{x}_{ij} = \sum_{k} a_{ik} f_{kj} \quad (19)$$

For example, the estimated amount of He for the first data set sample of the Corbigny area can be computed given the scores of Table 2 and the related loadings as:

He (estimated) = 0.6914 * (-7.1809) + 0.1229 * (30.3473) + 0.1856 (0.1836)

The difficulty arising here, is that considering strict mixing models, negative factor scores or negative factor loadings are not relevant. Thus, in situations where negative composition scores are unacceptable, they can be avoided by the selection of a different set of reference axes. Moreover, in any situation where either the addition of an end-member as indicated by a positive loading or the subtraction of an end-member as indicated by a negative loading, is unreasonable, a new selection of reference axes may be desirable. In fact, given the initial solution provided by the factor model, an infinity of rotated solutions can be computed. A simplified factor model given Equation (7), can be written in matrix form:

$$X_{(nxp)} = A_{(nxk)} F_{(kxp)} \quad (20)$$

considering a regular transformation matrix $T_{(kxk)}$, and its inverse T^{-1}, Equation (20) can be rewritten in the following form:

$$\hat{X}_{(nxp)} = (A_{(nxk)} T^{-1}_{(kxk)}) (T_{(kxk)} F_{(kxp)}) \quad (21)$$

thus leading to a new solution expressed as:

$$\hat{X}_{(nxp)} = A^{*}_{(nxk)} F^{*}_{(kxp)} \quad (22)$$

giving new factor scores and new factor loadings satisfying some constraints such as positive scores and loadings when modeling the mixing of various solutions. The estimated compositions, the residuals and the coefficients of determination (for the same number of factors) are thereafter the same for any model. The following transformation matrix was selected for the factor model shown in Table 3.

Table 3. Transformation matrix used to rotate initial solution displayed in Table 2, to respect positive constraint applying to scores and loadings in situation true mixing model

0.694	0.163	0.143
0.000	1.000	0.000
0.180	0.260	0.560

Applying Equation (21) to the data of Table 2 and Table 3, leads to the new scores presented in Table 4.

Table 4. Rotated factors scores for true mixing model of Corbigny area satisfying positive constraint

He	-0.010688	30.347321	6.700572
F-	0.602021	0.414385	-0.002989
HCO_3^-	67.207581	6.362858	16.281891
$SO4^{2-}$	-0.074349	5.954410	43.102112
Cl^-	4.458806	30.414734	7.835640
SiO2	4.178469	0.974594	-0.014434
Ca2+	16.733231	2.070524	16.202286
Mg2+	2.262041	0.694438	3.478939
Na+	3.217570	21.873993	6.457644
K+	1.425284	1.892729	1.958126

These positive factor scores now represent compositional end-members, and without providing further explanation of these results (it will be done in the related section), one should notice that each one of these new factor scores represents a different aquifer of the Corbigny area with its specific composition: the first factor corresponds to the silicified level (i.e. an intermediate geological layer), the second to the basement groundwater seepage with high He values, and the third to the superficial layer. Depending upon the local piezometric conditions, these various aquifers mix one another, and for each sample the contribution of each aquifer to the global measurements can be determined, the factor model providing a way to compute the proportions (i.e. loadings) of the different groundwaters involved. Given Equation (22), the factor model can be used to reconstruct the sample compositions, and to compute residuals representing the part of the geochemical signal not accounted for by the model. The factor model provides a relevant assessment of the background, whereas large-scale residuals should be considered as potentially indicating element enhancement (or depletion) which is not accounted for by the general trend for the data reconstruction achieved by the model. These residuals are of special interest when considering trace elements as they represent valuable candidates for the recognition of significant multi-elementary anomalies.

Once homogeneous subsets have been obtained and the modeling of compositional trends achieved, graphical projections of the samples in the factor space makes it possible to represent concisely the results of the model. The space dimension is determined by the number of factors kept for the model and the sample coordinates are merely the factor loadings. Usually, various projections are done using as reference axes the scores themselves, but noticing that:

$$\sum_{k} a_{ik} = 1 \tag{23}$$

for a three factor scores model, it is possible to represent the whole factor space as a single compositional plane (Fig. 1), all the samples lying in a restricted part of this plane in so far as loadings are positive and obey to the constraint expressed by Equation (23).

The representation displayed in Figure 1 also can be considered as a ternary diagram, the extremity of which are P1 P2 P3, a type of graphic geochemists are more accustomed to than binary projections (F1F2, F1F3, F2F3). The new axes are represented as the x and y lines. It has been used thoroughly in this work and seems to be a more efficient tool to visualize compositional sample suites than stereographic projection. The three original factors have the following coordinates in this representation:

$$x_1 = -\frac{\sqrt{2}}{2}\,;\, y_1 = -\frac{\sqrt{6}}{6}\,;\, x_2 = \frac{\sqrt{2}}{2}\,;\, y_2 = \frac{\sqrt{6}}{6}\,;\, x_3 = 0\,;\, y_3 = \frac{\sqrt{6}}{3} \tag{24}$$

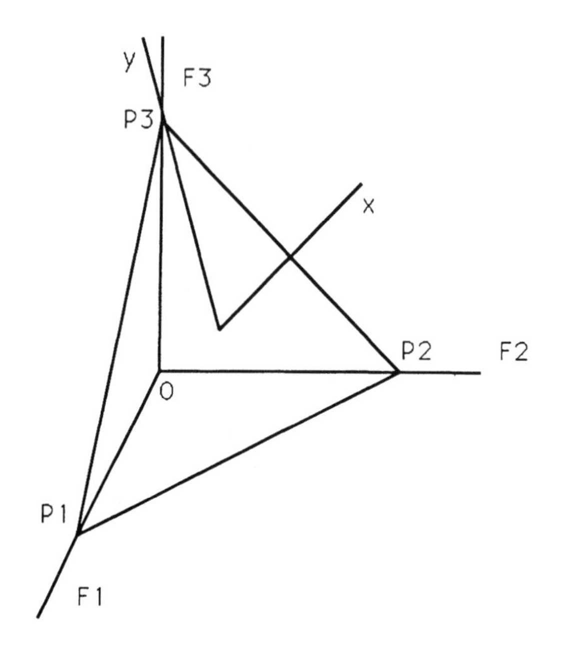

Figure 1. Compositional plane can be used to display all samples on same representation without requiring any projection, taking advantage of constraint expressed by Equation (22) and of positive property of compositional subspace (true mixing model).

and given any coordinate couple (x,y) observed in the plane P1 P2 P3, it is possible to compute its related loadings:

$$a_{i1} = \frac{1}{2} - \frac{\sqrt{2}}{2}x - \frac{\sqrt{6}}{6}y$$

$$a_{i2} = \frac{1}{2} + \frac{\sqrt{2}}{2}x - \frac{\sqrt{6}}{6}y \qquad (25)$$

$$a_{i3} = \frac{\sqrt{6}}{3}y$$

This representation makes it possible to study the entire compositional cloud on a single display, facilitates the selection of compositional end-members analyzing the structure of the factorial space, and to compute the loadings of the new poles given their related (x,y) coordinates in the P1 P2 P3 plane, thus leading to new end-members.

Once it has been possible to provide for a computational framework accounting for the mixing problem observed for the major constituents (estimating the end-members and their contributions to the sample measurements), a fundamental topic then is to reuse this model to explain the trace concentrations, characterizing the trace signal connected to each end-member and thereafter determining the source and regular amount of the trace elements. Regression analysis is a powerful technique to achieve such an objective, and residuals of this model represent significant anomalies, unexplained by any of the models. The Lodeve case study is an example where obvious wild values were meaningless and using careful methods, subdividing the original data set in various homogeneous subsets, confirming the clustering results with factor analysis also involved to account for the mixing model, then explaining trace concentrations thanks to regression analysis, made it possible to recognize extremely low values anomalies, connected to significant uranium ore bodies. Let us consider now the multiple regression model enabling us to estimate the background concentration for trace elements on the basis of the factor analysis results for the major constituents.

Regression Analysis

In this section we will consider multiple regression which is a further generalization of regression analysis to many-dimensional space. The variable being examined is the dependent or regressed variable designated Y_i and is modeled as a linear combination of independent variables or regressor variables denoted the X's. The general formulation of such a model is given by Equation (26):

$$\hat{Y} = \beta_0 + \beta_1 X_1 + \beta_2 X_2 + \ldots + \beta_m X_m + \varepsilon \qquad (26)$$

where e is a random variable of zero mean and s^2 variance, having a normal distribution. The criterion used to fit the regression is usually a least-squares solution, thus minimizing the following expression:

$$\sum_{i=1}^{n} (\hat{Y}_i - Y_i)^2 \qquad (27)$$

thus leading to the resolution of a set of normal equations which can be written in matrix form in the following way (Davis, 1973):

$$\begin{bmatrix} \sum X_0 & \sum X_1 & \sum X_2 & \cdots & \sum X_i \\ \sum X_1 & \sum X_1^2 & \sum X_1X_2 & \cdots & \sum X_1X_i \\ \sum X_2 & \sum X_2X_1 & \sum X_2^2 & \cdots & \sum X_2X_i \\ \cdot & \cdot & \cdot & \cdots & \cdot \\ \sum X_i & \sum X_iX_1 & \sum X_iX_2 & \cdots & \sum X_i^2 \end{bmatrix} \begin{bmatrix} \beta_0 \\ \beta_1 \\ \beta_2 \\ \cdot \\ \beta_i \end{bmatrix} = \begin{bmatrix} \sum Y \\ \sum X_1Y \\ \sum X_2Y \\ \cdot \\ \sum X_iY \end{bmatrix} \tag{28}$$

where X_0 is a dummy variable equal to 1 for each observation, thus the sum (on the X_0) being equal to n, the number of samples. The resolution of this set of normal equations provides the values of the coefficients of the regression bi. An estimation of the goodness-of-fit of the regression can be provided by the multiple correlation coefficient R, the ration of Equation (29) being near unity if the model is correct:

$$R = \sqrt{R^2} = \sqrt{\frac{SS_R}{SS_T}} \tag{29}$$

where SS_R is the sum of squares as a result of regression computed as:

$$SS_T = \sum_{i=1}^{n} (Y_i - \bar{Y})^2 \tag{30}$$

and where SS_T is the total sum of squares given by Equation (31):

$$SS_R = \sum_{i=1}^{n} (\hat{Y}_i - \bar{Y})^2 \tag{31}$$

The aim is to select a regressed variable, one of the trace elements to be studied, and to provide a model explaining the observed concentrations and modeling the background values for this element thanks to a combination of regressor variables. A good way to summarize the information on the major constituents is to reuse the results of the factor analysis, each factor accounting for a given aquifer in our model, and to consider the factor scores as the regressor variables. Trace concentrations will be modeled as a linear combination of the compositional water sources, and trace values then are explained directly by the mixing model, each aquifer providing some amount of the trace element. Considering the piezometric conditions, indirect prospecting (i.e. in the overburden basement) is straightforward when the basement water flows upwards through the surface water, thus providing an immediate indication of the basement characteristics. Residuals of the multiple regression model are an efficient way to recognize trace concentrations remaining unexplained by the mixing model, thus representing valuable prospects. The selection of the trace elements to be regressed is based on the correlation matrix computed for the traces and the factor scores, thus permitting to identify the trace elements showing a good relationship with the factors. An example of regression analysis for the He concentrations taking as regressor variables the factor scores displayed in Table 4, is provided for the Corbigny case study. Residuals as any other variable such as major constituents, traces, clusters, factor scores or complex subdomains defined by logical expressions (i.e. predicates expressed as conjunction of disjunctions of basic variables) or by expert-system rules can be mapped, providing direct insight into the spatial ordering and the spatial distribution of the elaborated criteria.

Mapping and Graphical Displays

Geochemists and hydrogeologists are well trained to use graphics, maps and many other specialized displays (Howarth, 1977, 1983; Howarth and others, 1980; Fisher, 1983), confirming the classical proverb: a picture is worth a thousand words. Moreover, many of the previous results can be mapped either using a discrete form or a regionalized approach interpolating (in some sense) values from regular grids, thus providing new data to be later integrated in the decisional process.

Therefore, four fundamental types of maps can be distinguished and handled by the system we developed:

- local value maps displaying square codified spots accounting for any type of symbolic representation. Classes coming from cluster analysis can be represented in such a way, selecting for each sample a color corresponding to its class and varying the saturation level (of the related color) according to distance of the sample to the class center;

- contour maps representing quantitative variables taken as basic measurements such as altitude, resistivity, remote soundings such as radiometry, or as model results such as factor scores;

- polygon maps showing classified or qualitative variables remaining constant within polygons, such as mining works, lithofacies, or corresponding to results elaborated using logical predicates (i.e. areas located at less than 20 meters from a leucogranite contact, for example);

- maps of classified lines such as faults, veins, lithological boundaries, or delimited by the results of image processing (i.e. contour detection, for example);

Moreover, external images also can be incorporated to this information system, mainly digital remote-sensing scenes either collected using airborne platforms or satellite scanners on various wavelengths such as the MSS channels of LANDSAT or stereographic couples for the Spot imagery system for example.

Most of the information gathered as basic geochemical variables (i.e. major or trace element concentrations in rocks, in water drillings or in stream sediments) or as physical measurements (i.e. such as resistivity, radiometry (Duval, 1976)), computed as model results and codified in the form of variables (i.e. one byte for each sample value) or of characteristics (i.e. one bit for each observation accounting for the boolean result of a logical predicate) (McCammon and others, 1983), or collected as images thanks to various sensors, can be handled in a geographic information system making it possible to build the databank. Homogeneous maps representing the spatial distribution of regionalized variables, are the results of numerous processing, interpolating contour maps between digitized contours to provide smooth models, estimating and interpolating concentrations between regular grid points computed from the initial random sampling (e.g. in the situation of geochemical or theoretical variables such as factor scores or residuals), or rescaling and updating different (or even incompatible) representation systems. Various techniques can be used to provide valuable estimators such as kriging (Matheron, 1972; Clark, 1977b), or to ensure interpolation between irregularly distributed observed or computed values at sample points such as trend-surface analysis (Clark, 1977a) or linear weighting. The maps we produced were computed mainly with the last method, thus yielding interpolated maps with a rustic estimator (compared to kriging) but making no serious assumptions on the variable properties such as stationary (for the kriging estimator) or polynomial distribution (in the situation of trend surface approximation), for example. Each node to be assessed is computed using the six nearest control points:

$$\bar{z} = \frac{\sum_{i=1}^{6} \frac{1}{\sqrt{d_i}} z_i}{\sum_{i=1}^{6} \frac{1}{\sqrt{d_i}}} \quad (32)$$

various linear combinations being later on possible, to enhance the influence of the nearest of the six control points, for example. Finally maps produced usually are transformed using spline functions to yield larger images increasing the legibility and to conform to the scale adopted for the remainder of the data bank documents. This map and image database summarizes all the information collected or modeled by the system and will represent the basis of further processing to define favorable characteristics during the data integration phase or to support the reasoning of expert modules.

Data Integration

Once this overall set of data has been collected and stored in an adequate form in the databank, the next logical step of the treatment process is to integrate various information layers into decisional maps (or parameters) accounting for favorable (respectively unfavorable) parts of a geochemical script (Mellinger and others, 1984; Leymarie and Durandau, 1985). Data integration follows the same reduction space logic as was involved in factor analysis but going one step further, trying to substitute to a large figure of variables (some of them being at this stage of high level such as model results) synthetic maps giving some insight into fragments of metallogenic scenarios. For each partial script accounting for some metallogenic factor, a predicate is built to account for the corresponding evidence in a rule-based reasoning framework. Training areas make it possible to assess the relevance of each scenario proposed based on a probabilist approach and in favorable situations lead to the inference of the corresponding hypothesis.

Let us briefly describe the formalism underlying this approach. We consider an *event* E_j related to a surface elementary unit, described by a logical *predicate* stating the validity of that event for the cell considered. Predicates are built using the databank maps or images and the logical and arithmetic operators excluding at this stage existential or universal quantifiers. Thus typical predicates are such as "the resistivity is less than 300 W m / m^2" or "select areas located at less than 20 meters from a redox barrier" and the *domain* D_j where a given predicate displays the logical value "true" is referred as the *extent* of E_j, i.e. the geographic area where the event occurs. This approach handles at the same time the statistical space thanks to the event based definition, the geographic space delimiting the extent of the statistical realization of E_j, and the logical space characterizing occurrences by logical predicates.

Complex predicates can be built, involving the well-known algebraic relational operators including conjunctions and disjunctions of elementary predicates. Following Leymarie and Heckert Gripp (1983), let us now define target indicators and objectives. Considering an event H, characterized by a predicate P and referred to as the objective, we are interested to know its status (true or nil). Let E_j, another event, be an indicator of the realization of H and let us consider:

- pr (H) the a priori probability of H;

- pr $(H \mid E_j)$ the conditional probability of H given the realization of E_j; the relevance of E_j as an indicator of H can be verified stating:

$$\frac{pr(H \mid E_j)}{pr(H)} \neq 1 \tag{33}$$

which simply indicates that the realization of the event E_j should modify the a priori probability of H. We may encounter favorable indicators (E_j reinforces the presence of H) or unfavorable ones (E_j expresses the absence of H).

$$\frac{\mathrm{pr}\left(\frac{H}{E_j}\right)}{\mathrm{pr}\,(H)} > 1 \tag{34}$$

Equation (34) expresses that E_j is an indicator of the presence of the objective H and:

$$\frac{\mathrm{pr}\left(\frac{H}{E_j}\right)}{\mathrm{pr}\,(H)} < 1 \tag{35}$$

Equation (35) expresses that E_j is an indicator of the absence of the objective H.

Noticing that:

$$\frac{\mathrm{pr}\left(\frac{H}{E_j}\right)}{\mathrm{pr}\,(H)} = \frac{\mathrm{pr}\left(\frac{E_j}{H}\right)}{\mathrm{pr}\,(E_j)} \tag{36}$$

it is possible to define the *reliability of the target indicator* as the probability of seeing the objective H realized when the indicator E_j occurs, that is pr (H / E_j), and the *detection rate* of the objective H by the indicator E_j, which is the probability of observing the indicator E_j when the objective H is satisfied. The prospecting strategy will aim to maximize one of these two incompatible criteria, thus increasing the detection rate if the objective is to inventory resources in a comprehensive way or the reliability if the goal is to avoid unsuccessful in-situ mining operations. Given training areas where it is possible to test the relevance of various indicators (some known mineralization exists), it is possible to calculate the respective probabilities required to compute the ratios of Equations (34) and (35) thanks to measurements of the frequencies of occurrences of the related events for the corresponding database images. The assessment of the a priori and conditional probabilities relies on the computation of the image histograms (pixel counting) for the logical expressions formed. Two limitations affect this approach, the first one comes from the necessity to rely on training areas seldom available in true prospecting and the second is the extremely step-by-step computational methodology involved. For each indicator, for each domain considered, for each objective addressed, the related probabilities have to be computed stepwise, storing logical masks and evaluating in an incremental way the logical predicates and their combinations (disjunctions of conjunctions of elementary predicates), storing the associated histograms and pixel counting.

But the main pitfalls arising from the system developed came from technological topics. Multivariate data processing is achieved by a knowledgeable engineer, preliminary interpretation and all mappings are achieved by hand, the constitution of the databank is time consuming and monitored by a human specialist, data integration supposes a trained engineer to imagine scenarios, to build the corresponding predicates and to test their relevance using training areas, then final interpretation of the results necessitates skilled experts of the processing techniques and of the geochemical processes involved. The conceptual approach is promising and significant results were obtained (Leymarie and Poyet, 1983; Poyet, 1986), but the methodology involved obviously is not realistic in terms of human resources required and efficiency reached. At the time this project was launched, no alternative was possible either from the software engineering viewpoint or from hardware limitations. Heterogeneous codes carrying out extremely diverse functions could not be integrated in a coherent and cooperative framework, each one being triggered according to the operator need, without a global automated logic which could be managed by a more intelligent software layer, able to achieve on its own the nondeterministic process of data analysis, to build the map and image corpus (for model results), to assert partial hypothetical metallogenic scenarios carrying out alone the data integration phase and finally to fit the scenarios together in a coherent attempt to provide a validation of some prospecting policy. Major progress had to be achieved to imagine an integrated software embedding the complete set of basic functionalities required for elementary processing needs, but moreover providing for a metarepresentation level enabling the software to "know" what it is made of and to get a perception of what its functions are useful for, to provide the environment with a high-level processing logic leading to multicooperative expert architectures. The work we accomplished to develop a new software workbench taking advantage of the most recent and advanced software engineering techniques (Poyet, 1990), lead to design on the theoretical and practical bases described to achieve the treatment of geochemical data, a new generation of tool. Some of the features of such an environment are highlighted in the next section.

Expert Systems

Expert systems have been born as real world applications of artificial intelligence research efforts, and the PROSPECTOR system (Duda, Gaschnig, and Hart, 1979) is one of the first operational and promising prototypes to be designed, thus contributing to an early introduction of these techniques into the geoscience community. Architectures of such softwares have evolved during this decade, following increasing hardware capabilities and the improvement of software engineering techniques. Workstations and microcomputers today offer more processing power and memory storage facilities on highly integrated machines than were available on expensive computers a few years ago. The first expert-system generation is currently moving towards transportable small machines (McCammon, 1986) and can even lead, thanks to sophisticated software architectures, to in-situ operational computerized assistants such as the HYDROLAB system (Poyet and Detay, 1990; Poyet and Detay, 1989a, 1989b, 1989c; Poyet and Detay, 1988a, 1988b), whereas larger computers provide enough power to design new expert system architectures addressing much more challenging topics such as knowledge-base distribution (Nii, 1986a, 1986b), cooperative multiexpert systems (Corby, 1987), coupled with object-oriented database facilities (Valduriez, 1987; Sibertin-Blanc, 1988), reusable software components, hardware independent high-level bitmap displays (Scheifler and Gettys, 1986; Devin, 1988) incorporating hypertext and hypermedia facilities (Bauer and Holz, 1989; Bensadoun, and others, 1989; Sandvad, 1989), and in a close future integration of image and parallel processing. But the main characteristic of second-generation expert systems is to handle deep knowledge, thus not only providing a superficial behavioral model stating that when a particular event occurs such action should be undertaken, but offering a deep model of the states of a given system and providing a meta representation of what the system knows and what its components are useful for and under which assumptions. This is a radical evolution and allied with current machine capabilities and many other software advances, most of them relying on object-based technologies (Boock, 1986; Cox, 1986; Moon, 1986; DeMichel and Gabriel, 1987; Bobrow and Kiczales, 1988; Cointe, 1988; Meyer, 1988; Roche and Laurent, 1989), it makes it possible to drawup the bases of the future systems on which we are working. Making an analogy with a database conceptual model (Briand and others, 1989), three main software levels can be identified. The physical layer corresponds to the basic entities handled by the software, such as computer codes (i.e. multivariate analysis functions (Anderson, 1958; Cooley and Lohnes, 1971)) or maps or images stored in the databank. The logical layer accounts for the meta knowledge representation facilities required to manage the former entities, thus providing for each of them an object model describing, for example in the situation of computing codes, their applicability conditions, their inputs and outputs, interpretation strategies of the results, ways to assess the robustness of the estimates, etc. The boundary separating the physical from the logical layer acts as a mirror where each low-level entity reflects in its corresponding high-level object endowed with its associated methods providing a manner to send them requests and to get adapted behaviors as in message-passing-based programing systems (Moon and Weinreb, 1981; Goldberg and Robson, 1983). This logical layer accounts for the deep understanding of what is going on in the software activity and can be used to carry on further rule-based reasoning on the object-layer, modeling the software constituents. The conceptual layer finally takes these objects as a reasoning basis to pursue the global strategies to be followed by the system when trying to assess favorable target areas according to some generic scenario. The analogy used has no exact physical reality in the implementation policies but provides for an adequate mental framework to tackle with the software architecture, which is better described as a cooperative aggregate of expert components, each one taking in charge some circumscribed task. Isolated packages already have been imagined using such an approach, designing for example intelligent front-end on comprehensive statistical packages (Gale, 1986). We will not give further details of these software developments, as they are not the central topic of this paper, but it is worthwhile to emphasize the great complexity of such unsupervised systems aiming to handle extremely diversified entities, to achieve on their own data analysis, mapping of relevant parameters, rescaling of the maps produced and ensuring their storage within the GIS, to carry out image analysis taking full advantage of the object based representation (e.g. a generic map is able to respond to a generic surface area message but not to a *slope* or *height* message specifically attached to topographic maps defined as a subtype of the map typological hierarchy), to carry on the test of partial metallogenic scenarios handling rule-based predicates modeling spatial reasoning, to assess finally the reliability of some global prospecting framework on the basis of tremendously diversified knowledge sources coming from structural geology (e.g. locating favorable situations such as synforms, structural barriers), petrology (e.g. analyzing differentiation rates or particular lithofacies), metallogeny (e.g. stable species), and hydrogeochemistry in the context of this study, thus connecting the numerical models describing the solution compositions with the hydrogeochemical processes such as redox barriers or solution-mineral equilibriums. All these knowledge sources have to cooperate in a coherent software architecture to account for efficient target recognition.

CASE STUDIES

Formulation of the Water Mixing Model

The objective is to achieve an indirect exploration of the basement uranium content thanks to hydrogeochemical prospecting at drilling sampling sites, to look for blind mineralization hidden by a superficial sedimentary layer. Given the mathematical model described previously, the aim is to use the compositional Q-mode factor analysis to provide a mixing model enabling us to describe the global composition observed at each sample site as a mixture of various end-members corresponding to a weighted sum of the contributing aquifers described by the factor scores with proportions given by the factor loadings. Given n water samples E_i (i=1,n), for which we have p ionic measurements I_j (j=1,p) for the major constituents, representing the global observations corresponding to the mixing of r water end-members T_k (k=1,r), and given c_{kj} the concentration measured in mg/l for the j th ion in the water type T_k, the dry content of this end-member T_k is given by Equation (37):

$$s_k = \sum_{j=1}^{p} c_{kj} \tag{37}$$

given Q_i the mass of E_i, and Q_{ik} the mass of T_k for the sample E_i, we have:

$$Q_i = \sum_{k=1}^{r} Q_{ik} \tag{38}$$

and the related proportion of T_k for the water sample E_i, referred as b_{ik} is given by Equation (39):

$$b_{ik} = \frac{Q_{ik}}{Q_i} \tag{39}$$

with the following property:

$$\sum_{k=1}^{r} b_{ik} = 1 \tag{40}$$

and the concentration, assessed in mg/l for the j th ion of sample E_i, referred x_{ij} is obviously given by Equation (41):

$$x_{ij} = \sum_{k=1}^{r} b_{ik} \cdot c_{jk} \tag{41}$$

The model used to reconstruct the measured concentration is the extended Q-mode factor analysis, operating on constant row-sums data, and permitting to account for the geochemical signal, given a reduced number of factors r' explaining the main part of the variance observed for the data set. If we refer to z'_{ij} as the transformed value of the estimate z_{ij} of x_{ij} given Equation (42):

$$z_{ij} = \frac{100 \, . \, z_{ij}}{s_i} \tag{42}$$

the factor model provides an estimate z'_{ij} of the transformed data (i.e. with constant row-sums equal to 100) given Equation (43):

$$z_{ij} = \sum_{k=1}^{r} a_{ik} \cdot f_{kj} \tag{43}$$

where a'_{ik} represents for each measurement the proportions of the water end-members T_k and where the f'_{kj} accounts for the transformed concentration (i.e. with constant row-sums equal to 100) of the j ions for the modeled water end-member T_k. The a'_{ik} and the f'_{kj} verify the usual properties of factor loadings and factor scores, that is:

$$\sum_{k=1}^{r} a_{ik} = 1 \; ; \sum_{j=1}^{p} f_{kj} = 100 \tag{44}$$

To provide an interpretation framework of the model results, it is necessary to transform the concentrations f'_{kj} of each end-member, multiplying it with a coefficient l_k, so that the concentrations $l_k \cdot f'_{kj}$ are as similar as possible to those of an observed water or of a saturated solution in equilibrium with known rocks, such as:

$$\sum_{j=1}^{p} \lambda_k \cdot f_{kj} = 100 \cdot \lambda_k = s_k \tag{45}$$

where s_k is the dry content. The mixing model then yields the estimates z_{ik}:

$$z_{ik} = \sum_{k=1}^{r} \left(\frac{s_i \cdot a_{ik}}{100 \cdot \lambda_k} \right) \cdot (\lambda_k \cdot f_{kj.}) \tag{46}$$

allowing an easy interpretation, based on the equivalence of the proportions b_{ik} of Equation (39) with the term given by Equation (47):

$$\left(\frac{s_i \cdot a_{ik}}{100 \cdot \lambda_k} \right) \tag{47}$$

moreover, identifying the concentration c_{jk} of the end-members (i.e. water types of Equation (37)), to the corresponding $l_k \cdot f'_{kj}$ of (46). It then is possible to characterize the composition of each water pole and to provide a quantitative assessment of the proportions of each estimated water type T'_k for each sample measurement. This interpretation framework will be reused throughout the case studies carried out to illustrate the relevance of this modeling approach.

Corbigny Case Study

The first site is located in the West part of the Morvan granitic horst, at the southwest of the volcano-tectonic *Montreuillon* basin. Granites, cut by great N-S faults are covered by Mesozoic limestones (Fig. 2). The objective is to locate uranium hidden deposits under a covering sedimentary layer ranging from 10 to 100 meters thick. The drillings carried out provided rock geochemistry and hydrogeochemistry data including for the latter the piezometric surface observation, the transmissivity, pH, major element concentrations, and values for U, Rn and He. At the interface between the different layers, groundwaters flow on an aquifer of about 10 km^2. Their leaching on granitic rocks along microfissurations (especially for the boundary separating the basement from the silicified layer), is the key of this type of prospecting.

Figure 2 provides a good insight into the geological context fairly well summarized by the opposition between the basement layer and the recent sedimentary formations. The basement can be subdivided into three main components:

- a volcanic occurrence at the north, itself subdivided into the ignimbritic *complexe de Blismes*, a carboniferous layer of 50-meters thick which

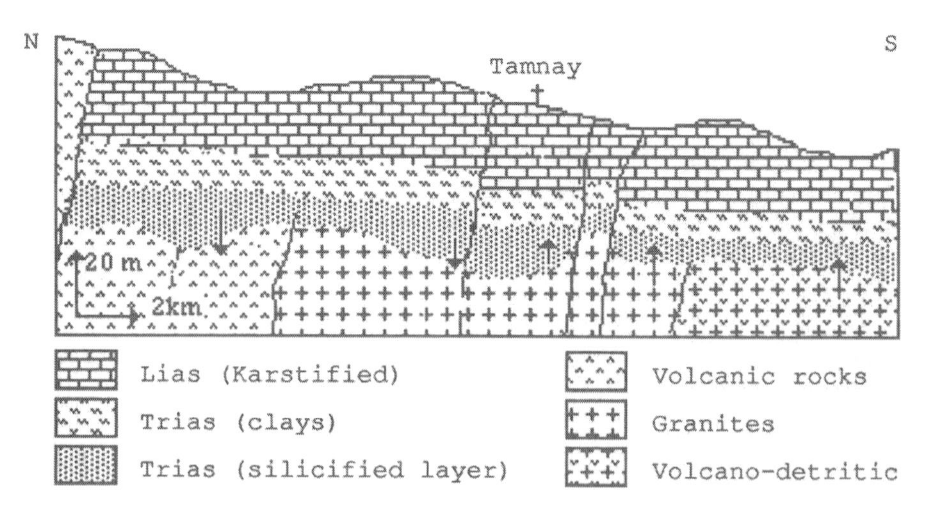

Figure 2. Geological context of Corbigny area.

can be eroded locally and the *complexe de Montreuillon*, representing the upper part of this volcanic layer;

- foliated microgranites observed all around the caldera and intrusive in the borders;

- granites of *Château Chinon* representing the infrastructure of the basin, characterized as phorphyroid monzo-granites;

This basement can be considered from a hydrogeological point of view as a rather homogeneous unit, corresponding to a unique aquifer polluted by descending waters from the surface in the northern part of the area, with a generalized directional water flow oriented towards the south, according to the piezometric surface. This basement aquifer (referred to as unit 1), finally percolates through the surface groundwaters when the hydraulic charge in the southern part of the studied area is sufficient. The sedimentary cover includes a Triassic silicified layer, topped by Triassic fine-grained clayey layers, the sequence being ended by a karstified Lias. The hydrogeological framework is characterized by an aquifer hosted by the silicified layer (i.e. unit 2), isolated from the upper groundwater seepage (karstified aquifer or unit 3), thanks to the Triassic marls. As we mentioned briefly, communications between units 3 and 2 are the result of major tectonic features such as faults and fractures, and those between units 2 and 1 are regulated by piezometric gradients, leading to the contamination of the basement groundwaters by surface flows in the north and corresponding to a reverse scheme in the south where the basement aquifer percolates through the silicified layer. The location where flows reverse between units 1 and 2, can be situated near to the *Tamnay-en-Bazois* village (Fig. 2).

According to this complex hydrogeological framework and to the goals followed (i.e. indirect prospecting of the basement), our aim is twofold; first recognizing outliers with a good confidence (to study them apart), then obtaining a mixing model enabling us to map the various water compositions and to compute the mixing proportions between the three groundwater units for each sample analyzed. In fact we provide here a summary of the real approach followed, more complex indeed when considering the operational study, and describing a simplistic reasoning scheme. The central point to be kept in mind is that we cross-check results from various methods to obtain robust analysis, modeling first major compositions and trying then to take advantage of the results to account for trace content.

The first step is to compute many cluster analyses of the initial data set based on the composition of the major constituents, each run providing results such as those reported in Table 1. Successive runs lead to rather similar outcomes making it possible to recognize stable compositional water classes outlining a topological ordering in the classification space, and to point out remote samples from the model. Representing the robust classes of the classification space using a ternary compositional diagram makes it possible to display a structural ordering. Figure 3 moreover leads us to suspect for a mixing model between three main end-members, intermediate classes accounting either for local variations or for transitional compositions.

The distribution in the compositional space of the robust classes identified (Fig. 3), lead us to suspect a mixing model between compositional end-members represented by the sulphated waters of the covering (i.e. represented by class 1), the bicarbonated calciferous waters of the silicified layer (i.e. represented by classes 2 and 4) and the chloridized and sodic waters coming from the basement (i.e. represented by class 7).

Moreover, for each run, samples are classified as in Table 5, the labeling done permitting us to assess the typicality of each measurement. Finally, samples suspected to represent outliers for the major elements model and the mixing model itself must be cross-checked using another method, such as factor analysis for example.

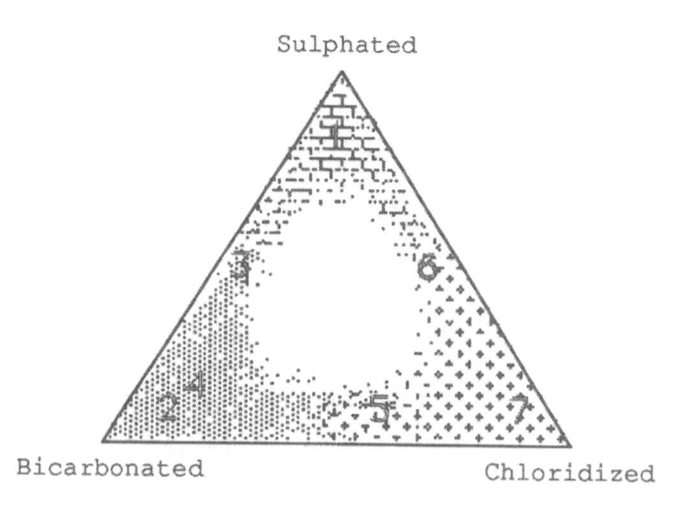

Figure 3. Structural ordering of classes in compositional space (same patterns as for Fig. 2).

Table 5. Sample classes: each sample is labeled with its residual inertia (ratio between distance to gravity center to intrainertia of population)

Six elements belong to class 1:					
BA03(170)	BA16(299)	BA18(456)	AU34(557)	AU33(1939)	BA19(3739)
Twelve elements belong to class 2:					
BA21(5)	AU15(15)	AU14(16)	BA01(18)	BA10(21)	AU03(22)
BA12(24)	BA22(35)	BA13(35)	BA15(46)	AU10(81)	AU12(170)
Five elements belong to class 3:					
AU21(15)	AU09(31)	AU36(134)	AU08(261)	AU11(497)	
Twenty-one elements belong to class 4:					
BA05(1)	AU27(2)	AU26(4)	AU25(4)	BA20(4)	AU24(7)
AU01(9)	AU04(10)	AU02(13)	BA04(13)	BA17(14)	AU20(18)
AU38(21)	BA11(22)	AU13(28)	AU06(32)	BA06(36)	AU16(42)
AU17(62)	BA02(65)	CR05(116)			
Five elements belong to class 5:					
AU40(5)	AU41(6)	AU35(40)	AU31(145)	AU28(267)	
One element belongs to class 6:					
AU19()					
Eight elements belong to class 7:					
SPCH(17)	AU32(28)	AU30(43)	AU39(54)	BA34(60)	BA08(63)
BA07(73)	AU29(91)				

An immediate advantage of the classification techniques is to lead to a straightforward mapping of the compositional classes, assigning a specific color to each class and saturating it according to the distance of each sample to its class center, providing in this way for discrete maps at sample sites (i.e. local value maps). Then robust shapes can be identified (i.e. collecting samples being frequently classified together) and also mapped to assess the spatial distribution of the groupings given by robust clusters. The most interesting property arising from classification techniques and justifying their use as a preliminary approach to the data-set structure stems from the fact that the results obtained are close to the initial data and once an interpretative framework has been elaborated, it is possible to visualize the geological reality without distorting it with complex models. Of course, once the preliminary investigation has been made, it is worthwhile to compute a continuous model, confirming the supposed outliers, distributing the measured concentrations between the various end-members recognized and permitting contour maps to be drawn. This is the purpose of Q-mode factor analysis.

For this case study, and based on the values of the coefficients of determination (i.e. factor-variance diagram), a three factor model accounts well for all variables except K, Mg, and F. A fourth factor would explain moreover Mg and F, but for the sake of simplicity the three factor model is deemed as satisfactory. Initial results of the factor analysis were reported in Table 2, then rotated using the matrix of Table 3, thus leading to the positive compositional end-members of Table 4. A brief analysis of these results provides a confirmation of the mixing model previously supposed (Fig. 3), factor 1 accounting for the composition of the silicified groundwater level, factor 2 modeling the composition of the basement aquifer, and factor 3 representing the surface waters of the sedimentary covering.

Using the representation previously described to display all compositional variations in the same P1-P2-P3 plane, it is possible to obtain Figure 4. The result of the computation, expressed in this graphical way, is an interesting mixing model showing the groundwaters of the silicified aquifer as the starting point of two geochemical trends joining the basement waters and the surface aquifer. Moreover it provides an efficient way to identify outliers standing outside of the main geochemical trends, characterized by high residual values and by contrasted factor loadings (i.e. unrealistic proportions).

The calciferous and bicarbonated water-pole of the silicified aquifer is represented by the factor 1 of the factor model and its composition can be summarized by the two following representative samples for this pole (Table 6):

Table 6. Characteristic samples for silicified aquifer

	He	F-	HCO3	SO4	Cl-	SiO2	Ca2+	Mg2+	Na+	K+
BA21	.27	.39	360.87	20.40	18.44	20.00	106.29	8.46	5.75	5.0
AU14	.16	.62	319.74	13.75	21.27	11.60	111.10	0.68	4.90	3.0

This groundwater end-member is polluted, according to the local piezometric conditions, by the surface water in the northern part of the area studied and by the basement groundwaters in the south. The composition of the chloric and sodic water of the basement, enriched in Helium, located on the bottom right part of Figure 4 and modeled by factor 2, is well summarized by the three following samples:

Table 7. Characteristic samples for basement groundwaters

	He	F-	HCO3	SO4	Cl-	SiO2	Ca2+	Mg2+	Na+	K+
BA07	355.27	3.50	234.56	75.20	241.11	12.80	57.31	11.04	249.00	8.20
BA08	373.41	4.20	236.88	92.80	376.55	8.50	61.56	9.53	310.00	12.45
AU29	358.23	3.00	228.32	108.00	429.03	18.75	83.37	12.74	284.00	10.65

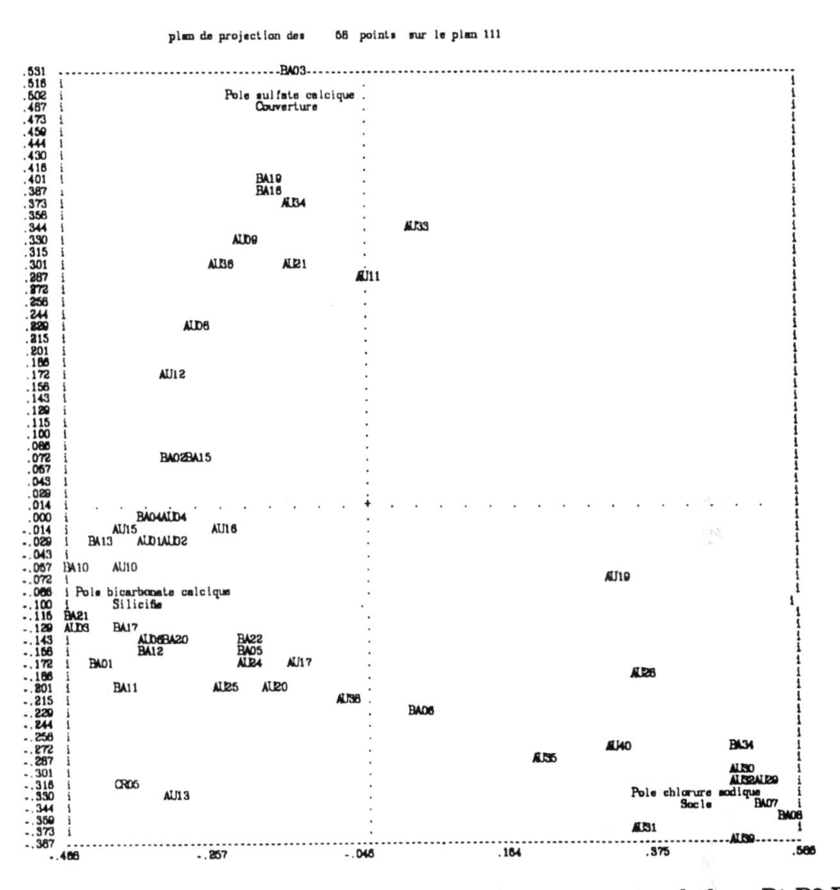

Figure 4. Results of mixing model represented in compositional plane P1 P2 P3.

Finally, characteristic composition for the calcic and sulphated waters of the surface appearing on the top of Figure 4 and represented by factor 3, can be described given two representative samples such as Table 8:

Table 8. Characteristic samples for surface aquifer

	He	F-	HCO3	SO4	Cl-	SiO2	Ca2+	Mg2+	Na+	K+
BA18	0.21	.24	293.00	1400.0	8.50	1.00	535.00	60.00	7.50	10.00
BA03	1.13	.02	269.22	1190.0	27.66	12.20	419.08	22.47	9.20	4.60

The composition of the samples located on Figure 4 between the mentioned hydrogeochemical end-members and constituting the observed trends can be explained given the mixing model as a blending of the three poles according to the proportions described by the factor loadings. For example, the composition of each one of the samples is reconstructed as the mixing of the three poles according to the loadings of Table 9.

Table 9. Loadings for five first samples of Corbigny data set

BA01	.9749	-.0574	.0825
BA02	.6003	-.1154	.5151
BA03	-.0624	-.3063	1.3707
BA04	.7176	-.1165	.3992
BA05	.7223	.1271	.1507
etc ...			

Residuals can be computed for the mixing model and represent the unexplained part of the geochemical signal for the major constituents. For example Table 10 shows the residuals for the five first samples of the data set:

Table 10. Residuals of mixing model for major constituents

BA01	1.54	-.44	3.34	-2.59	-1.30	-1.59	-1.50	- .73	.19	3.09
BA02	.09	-.30	5.10	-5.67	2.12	-1.01	.04	.25	-.60	-.02
BA03	.23	.17	-2.39	3.60	.32	1.20	.90	-3.28	-.07	-.71
BA04	.94	-.27	2.79	-3.80	1.84	-1.22	1.41	- .61	-.59	-.28
BA05	-.43	.34	- .52	2.85	-1.35	-1.00	- .43	.07	.49	-.03

Samples presenting high values, and not located within the geochemical sequences or within the pole themselves, should be rejected as outliers when the cross-checking with the results obtained from the clustering is positive. They are studied apart and the trace thresholds are defined specifically for these particular populations (they present no interesting features for this Corbigny case study). For the general population, the framework involved in studying the trace compositions is twofold: files are sorted according to the clustering achieved for the major constituents and the subpopulations trace contents are characterized. Then significant traces which can be used as uranium tracers (i.e. such as direct tracers like Rn or He or indirect ones as F^- (mineralization is associated to fluorine in the Corbigny area)) and showing significant correlation with the mixing model developed for the major constituents are regressed (i.e. factors are used as regressor variables) to compute the geochemical background.

Then residuals are substituted to the observed values when possible, and a classification for the "trace signal" can be achieved highlighting sensitive areas representing valuable prospects. Finally according to the hydrodynamic scheme and depending upon the hydrogeochemical context (i.e. equilibrium conditions according to pH and Eh diagrams) it is possible to provide for a global conceptual interpretation of the observed phenomenons. Let us describe briefly results obtained for the Corbigny area following that framework.

The selective study of the trace content for the geochemical classes computed for the major components and illustrated by Table 1 shows that surface waters (i.e. classes 1 and 3) present no particular anomaly of the trace signal except that their content seems to be diluted. Basement groundwaters (i.e. classes 5, 6, and 7) correspond to measurements made in the southern part of the studied area where they can percolate through the upper layers and show high values of the He and F- content. An intermediate composition represented by class 4, corresponding to the mixing of the silicified aquifer with the basement groundwaters (in a location where the piezometry begins to allow it) shows the most interesting multiple anomalies both for Rn and for the U dry residue, and samples AU25, AU26, AU38, BA06, AU17, CR05 grouping geographically together already seem to be most interesting.

Keeping in mind the general properties of the three main aquifers for the trace content, it now is possible to try to explain the trace values thanks to the former mixing model developed for the major components. Of course a preliminary check is to verify that a close correlation exists between the traces and the factor scores. Table 11 provides a summary of the correlations computed:

Table 11. Correlation matrix for traces and factor scores (mixing model for major components) - U-w stands for Uranium content in water and U-dr for Uranium dry residue

	Rn	He	U-w	U-dc	F-	F1	F2	F3
Rn	1.00	-.17	.30	.58	.31	.13	-.03	-.23
He		1.00	-.09	-.14	.43	-.37	.91	-.26
U-w			1.00	.54	.08	-.04	-.25	.41
U-dr				1.00	.34	.00	-.03	-.19
F-					1.00	-.18	.54	-.40
F1						1.00	-.16	-.36
F2							1.00	-.63
F3								1.00

Three trace elements can be supposed to obey the mixing model developed for the major components, that is He, U (water content), and F-. For these elements, we are going to compute regressions and residuals representing finally significant anomalies not accounted for by the mixing model. The first regression analysis computed accounts for the He content. The following coefficients of regression have been obtained for the least-square solution under constraints (Table 12):

Table 12. Coefficients computed for regression of He taking factor scores as regressors

Name	Coef	Sigma	t Student	Coefficient	Sigma (norm.)
F1	.641	.812	.789	.0150	.0190
F2	29.799	.545	54.669	1.0994	.0201
F3	6.303	.252	24.880	.5239	.0210

Contribution of the various factors seems to be as a confirmation of the hydrogeochemical hypothesis made concerning the He origin (refer to the characterization of the trace content of classes 5, 6, and 7). Factor 2, corresponding to the basement groundwater proportion shows the higher contribution (refer to coefficient values) thus proving the basement origin of this element, representing a valuable indicator of the U rock content (only in the southern part of the area). A straightforward evaluation of the goodness-of-fit of the regression model can be displayed graphically, taking as an abscissa the trace values and representing on the other axe the estimated values. The straightline displayed in Figure 5 is then an indicator of the correctness of the model.

The graph shows that the hypothesis of the basement origin of the He content is satisfactory. It then is possible to substitute for the measured values of He the residuals computed by that regression model, keeping only the unexplained part of the geochemical signal. Regressions made for U and F- did not seem to be satisfactory and these elements do not conform to the mixing model developed for the major constituents. Therefore, trace classes are computed given Rn, U (water content), U (dry residue), F-, and regression residuals for He representing the regionalized anomaly for that element. Results for such a model are displayed in Table 13.

This analysis shows two diluted water classes corresponding to the silicified aquifer (classes 1 and 2), a class showing a depleted content of

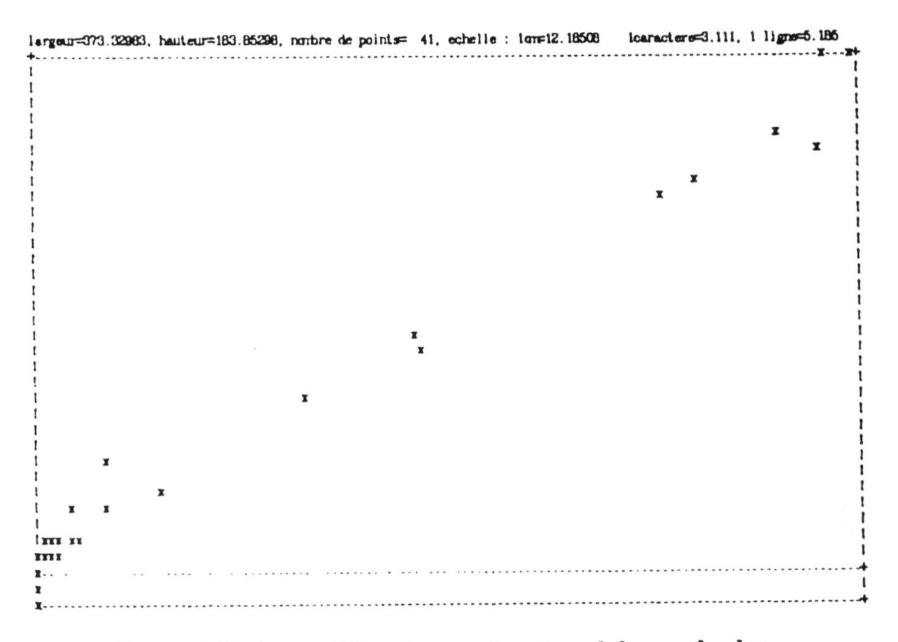

Figure 5. Estimated He values as function of observed value.

Table 13. Cluster analysis for trace content and He residuals

Var	Pop (41)	Class1 (6)	Class2 (17)	Class3 (1)	Class4 (5)	Class5 (5)	Class6 (4)	Class7 (3)
Rn	22.9	31.2	11.8	22.0	11.4	7.8	37.9	93.2
U-w	2.5	1.8	1.5	4.8	4.4	1.5	4.2	4.7
U-dr	4.6	1.7	2.9	7.5	2.5	2.7	15.1	12.4
F-	2.2	1.1	1.8	3.5	1.6	3.3	2.8	4.5
He-res	-15.4	-14.6	-15.4	55.0	-56.5	14.5	-20.7	-15.3

He and a slight positive anomaly of U (water-content), two classes enriched in He (classes 3 and 5) and corresponding to samples located in the south part of the studied area characterized by the basement groundwater composition, and two classes depleted in He (classes 6 and 7) but showing relevant multielement significant anomalies enriched in Rn, U (water content), U (dry residue), and F-, located in the middle part of the area, at the north of *Tamnay-en-Bazois* (Fig. 2).

Let us consider now the drawbacks arising from this case study. First of all let us remember that the silicified aquifer considered in great detail in this study (the sampling available for this layer was more comprehensive than for any other aquifer) can provide valuable information about the rock content (this is the final objective of this indirect prospecting of blind orebodies) only in the southern part of the area, where the basement groundwaters can percolate through the upper layers due to the hydraulic charge. According to this observation, indirect prospecting of the basement should be possible only there, but one should notice that the mixing model developed shows that some mixing of units 1 and 2 also occurs in the central area, to the immediate North of *Tamnay-en-Bazois*. The interpretative framework is that two main anomalies have been recognized: the first one corresponds to high values for all traces studied (Rn, U-wc, U-dr, F-) and depleted values for He (residual), occurring in the central area of the zone studied, to the immediate North of *Tamnay-en-Bazois*, just where the mixing model shows slight pollution of the silicified aquifer by the basement groundwaters, and the second anomaly corresponds to high contributions of the basement aquifer in the extreme south part of the area (He enriched anomaly), where the global concentrations mainly are the result of the basement groundwaters signature. The conceptual interpretation of such distribution of the anomalies outlined is that a target has been recognized at the immediate North of *Tamnay-en-Bazois* and that the basement groundwaters, leaching this anomaly, flow towards the south according the general piezometric surface with a high He enriched content, and finally percolate through the surface aquifers, mainly polluting the silicified layer. This assumption relies on rock geochemistry, which led to the discovery of a blind orebody of limited extent (it is not a paying ore under the current economic conditions), emphasizing the sensitivity of the methods presented in this paper.

CONCLUSIONS

This work and the approach presented here to process the hydrogeochemical data gathered in the context of blind uranium exploration were inspired by Tukey's well-known book on *Exploratory Data Analysis*, where the author emphasized the importance of the following four facets:

- Graphical displays to reveal the structure of the data;

- Transformation and reexpression of the data using ad-hoc functions such as logarithmic or power transformations to simplify the data response and behavior;

- Robust (i.e. in the sense of statistical resistance) analysis methods ensuring that remaining outliers should not unduly distort a whole modeling attempt;

- Residual-based decisions, focusing the exploration geochemist attention on the unexplained remaining part of the geochemical signal accounted for by the models.

The root of this work is founded in such an approach, but was tailored for principles of operational practice in the specific domain of hydrogeochemical exploration of trace elements based on indirect ground-water-based methodology (thanks to sampling drillholes), and of course a complete framework was developed to gather, process, integrate, and represent geochemical data emphasizing some of the key-points of Tukey's approach (Tukey, 1977), developing new techniques, giving original interpretations to well-known ones, moreover providing a complete methodology tested on real-world programs. But the problem outlined in this paper is a difficult one as the data gathered represent an indirect appraisal of the rock content and as the values measured correspond to an overall bulk with no indication of the origin of its chemical content. Subdividing the samplings into various packages thanks to clustering techniques, and cross-checking the results using factor analysis lead to homogeneous composition poles corresponding to hydrogeological units mixing according to the hydrodynamic conditions prevailing in the area. These compositional end-members can be confirmed easily using factor analysis, and moreover a mixing model can be proposed to assess for each sample the mixing proportions of the main elementary compositional units contributing to the global measurement. Once this mixing model can be considered as satisfactory to explain the water content of the major constituents, factor scores can be reused as regressor variables to build a linear model accounting for the content of trace elements. Thus, mathematical techniques provide a powerful tool to understand what goes on in the data set, providing a reliable way to distribute samples within homogeneous subsets for which valuable modeling can be proposed, focusing attention on model residuals generally corresponding to an unexplained part of the geochemical signal. Considering trace elements properly accounted for by the regression analysis, residuals can be computed and used to characterize the regional anomaly much more significantly than the selection of unrelevant outliers correctly explained by the model. Of course, multielement anomalies as observed in this Corbigny survey should be considered as highly significant and should be later included in the databank after adequate mapping has been realized to achieve the data integration phase. Considering this didactic study, there was no need to carry out a time consuming integration process; nevertheless the complete methodology as presented has been extensively used to handle difficult situations requiring the assessment and validation of metallogenic scenarios involving extremely diversified data. In the current example, thanks to the aforementioned results and taking into account the hydrodynamic conditions of the area studied, an easy confirmation of the ore body was given revealing the sensitivity of such indirect groundwater-based prospecting techniques.

ACKNOWLEDGMENTS

I am indebted to Pierre Leymarie for the many suggestions and constructive criticism he provided throughout the various stages of development of this work. I am grateful also to Richard Sinding-Larsen for his advice and contribution to the examiner board of the D.Sc. dissertation this work was presented to, and to Alfred T. Miesch for the availability of various programs we tailored to the hydrogeochemical prospecting needs. Contributions of members of the COGEODATA group of the IUGS also should be acknowledged as should be the assistance of many colleagues providing valuable help in software development.

REFERENCES

Anderson, T. W., 1958, An introduction to multivariate statistical analysis: John Wiley & Sons, New York, 374 p.

Bauer, J., and Holz, D., 1989, Otello, An object-oriented implementation of an electronic book: Proc. Workshop on Object Oriented Document Manipulation BIGRE no. 63-64, p. 42-51.

Bensadoun, O., Chrisment, C., Pujolle, G., and Zurfluh, G., 1989, Bases d'informations généralisées, approche hypertexte pour la consultation de documents multimédia: Modèles et Bases de données, no. 13/14: p. 37-52.

Benzecri, J. P., and Benzecri, F., 1980, Pratique de l'analyse des données: l'analyse des correspondances: exposé élémentaire: v. 1, Dunod, Paris, 424 p.

Boock, G., 1986, Object-Oriented Development: IEEE Trans. on Software Engineering, v. SE-12, no. 2, 211-221.
Bobrow, D. G., and Kiczales, G., 1988, The common lisp object system metaobject kernel, a status report: Proc. International Workshop on LISP Evolution and Standardization, Paris, France, p. 27-32.
Botbol, J. M., Sinding-Larsen, R., McCammon, R. B., and Gott, C. B., 1978, A regionalized multivariate approach to target selection in geochemical exploration: Econ. Geology, v. 73, no. 4, p. 534-546.
Briand, H., Ducateau, C., Hebrail, Y., Herin, D., Kouloumdjian, J., 1989, Une démarche générale de conception d'une base de données du niveau logique au niveau physique: Modèles et Bases de Données, no. 12, p. 65-81.
Clark, I., 1977a, Snark, a four dimensional trend-surface computer program: Computers & Geosciences, v. 3, no. 3, p. 283-308.
Clark, I., 1977b, Practical kriging in three dimensions: Computers & Geosciences, v. 3, no. 2, p. 173-180.
Cointe, P., 1988: Towards the design of a CLOS Metaobject Kernel: ObjVlisp as a first layer: Proc. International Workshop on LISP Evolution and Standardization, Paris, France, p. 33-40.
Cooley, W. W., and Lohnes, P. R., 1971: Multivariate data analysis: John Wiley & Sons, New York, 364 p.
Corby, O., 1987: BB1 en SMECI: Proc. 6ème Congrès Reconnaissance des Formes et Intelligence Artificielle de l'Association Française de Cybernétique Economique et Technique, p. 581-586.
Cormack, R. M., 1971, A review of classification: Jour. Roy. Stat. Soc., London, Ser. A, v. 134, 321-367.
Cox, B. J., 1986, Object-oriented programming: An evolutionary approach, Addison-Wesley Publ. Co., Reading, Massachusetts, 217 p.
Davis, J. C., 1973, Statistics and data analysis in geology: John Wiley & Sons, New York, 550 p.
DeMichiel, L. G., and Gabriel R. P., 1987, The common lisp object system: Proc. European Conference on Object-Oriented Programming, (reprinted as Bigre+Globule, no. 54), Paris, France, 201-220.
Devin, M., 1988, AIDA: machine independent window/mouse interface development tool for Le-Lisp: Proc. International Workshop on LISP Evolution and Standardization, Paris, France, p. 55-61.
Diday, E., 1975, Classification automatique séquentielle pour grands tableaux: Jour. RAIRO, no. 3, p. 29-75.
Duda, R. O., Gaschnig, J. G., and Hart, P., 1979, Model design in the PROSPECTOR consultant system for mineral exploration, *in* Michie, D., ed., Expert Systems in the Microelectronic Age, Edinburgh Univ. Press, p. 153-167.

Duda, R., and Hart, P., 1973, Pattern classification and scene analysis: Wiley Interscience Publ., New York, 482 p.

Duval, J. S., 1976, Statistical interpretation of airborne gamma-ray spectroscopic data using factor analysis: Proc. Exploration for Uranium ore Deposits, IAEA, Vienna, Austria, p. 71-80.

Everitt, B., 1974, Cluster analysis: Heinemann, London, 122 p.

Fisher, N. I., 1983, Graphical methods in nonparametric statistics: a review and annotated bibliography: Intern. Stat. Rev., v. 51, no. 1, p. 25-58.

Gale, W. A., ed., 1986. Artificial intelligence and statistics: Addison-Wesley Publ. Co., Reading, Massachusetts, 418 p.

Goldberg, A., and Robson, D., 1983, Smalltalk-80, The language and its implementation: Addison-Wesley Publ. Co., Reading, Massachusetts, 234 p.

Harman, H. H., 1967, Modern factor analysis (2nd ed.) : Univ. of Chicago Press, 474 p.

Hoffman, S. J., and Mitchell, G. G., 1984, The micro-computers in geochemical exploration: Jour. Geochem. Explor., v. 21, no. 4, p. 437-454.

Howarth, R. J., 1977, Cartography in geochemical exploration: Sci. Terre Inform. Geol., no. 9, p. 105-28.

Howarth, R. J., 1983, Statistics and data analysis in geochemical prospecting: *in* Howarth, R. J., ed., Elsevier Science Publ., 437 p.

Howarth, R. J., 1984, Statistical applications in geochemical prospecting: a survey of recent developments: Jour. Geochem. Explor., v. 22, no. 1, p. 41-61.

Howarth, R. J., Koch, G. S., Chork, C. Y., Carpenter, R. H., and Schuenemeyer, J. Y., 1980, Statistical map analysis techniques applied to regional distribution of uranium in stream sediment samples from the southestern United States for the national Uranium Resource Evaluation Program: Jour. Math. Geology, v. 12, no. 6, p. 339-366.

Howarth, R. J., and Sinding-Larsen, R., 1983, Multivariate analysis, *in* Howarth, R. J., ed., Handbook of exploration geochemistry, v. 2: Statistics and data analysis in geochemical prospecting: Elsevier Science Publ., Amsterdam, p. 207-289.

Klovan, J. E., 1981, A generalization of extended Q-mode factor analysis to matrices with variable row-sums: Jour. Math. Geology, v. 13, no. 3, p. 217-224.

Klovan, J. E., and Imbrie, J., 1971, An algorithm and FORTRAN IV program for large scale Q-mode factor analysis and calculation of factor scores: Jour. Math. Geology, v. 3, no. 1, p. 61-77.

Klovan, J. E., and Miesch, A. T., 1976, Extended Cabfac and Q-model computer programs for Q-mode factor analysis of compositional data: Computers & Geosciences, v. 1, no. 3, p. 161-178.

Leymarie, P., and Durandau, A., 1985, The integration of geochemical exploration data, *in* Glaeser, P., ed., The role of data in scientific progress: Elsevier Science Publ., Amsterdam, p. 33-37.

Leymarie, P., and Heckert Gripp, A., 1983, Mise au point, documentation et évaluation d'un logiciel de traitement interactif des données géochimiques sur console image en couleurs: unpubl. report CNRS/DGRST, 220 p.

Leymarie, P., and Poyet, P., 1983, Géochimie en roche: relations avec les potentialités métallogéniques de l'uranium: Proc. Fifth Meeting of the Contact Group Uranium Exploration Methods and Techniques, European Community Commission, Brussels, Nov. 24-25, 5 p.

Matheron, G., 1972, Théorie des variables régionalisées, *in* Traité d'informatique géologique: Masson Publ., Paris, p. 307-378.

McCammon, R. B., 1986, The PROSPECTOR mineral consultant system: a micro-computer based expert system for regional resource evaluation: U.S. Geology Survey Bull. 1967, 35 p.

McCammon, R. B., Botbol, J. M., Sinding-Larsen, R., and Bowen, R. W., 1983, Characteristic analysis: Final program and a possible discovery: Jour. Math. Geology, v. 15, no. 1, p. 59-83.

Mellinger, M., Chork, S. C., Dijkstra, S., Esbensen, K. H., Kurlz, H., Lindqvist, L., Saheurs, J. P., Schermann, O., Siewers, U., and Westerberg, K., 1984, The multivariate chemical space, and the integration of the chemical, geographical, and geophysical spaces: Proc. Intern. Geoch. Explor. Symposium, p. 143-148.

Meyer, B., 1988, Object-oriented software construction: Prentice Hall International Series in Computer Sciences, Cambridge Univ. Press, 534 p.

Miesch, A. T., 1976a, Interactive computer programs for petrologic modeling with extended Q-mode factor analysis: Computers & Geosciences, v. 2, no. 4, p. 439-492.

Miesch, A. T., 1976b, Q-mode factor analysis of compositional data: Computers & Geosciences, v. 1, no. 3, p. 147-159.

Moon, D., and Weinreb, D., 1981, Flavours: Message passing in the Lisp Machine: MIT AI Lab. AI Memo no. 602 (1980), and the Lisp Machine Manual, Massachussets, p. 63.

Moon, D., 1986, Object oriented programming with Flavors: Proc. OOPSLA'87, Special Issue of SIGPLAN Notices, Portland Oregon, v. 21, no. 11, p. 1-16.

Nii, P., 1986a: Blackboard systems: the blackboard model of problem solving and the evolution of blackboard architectures: A.I. magazine, July 1986, p. 38-53.

Nii, P., 1986b. Blackboard system: blackboard application system, blackboard systems from a knowledge engineering perspective: A.I. magazine, August 1986, p. 82-106.

Poyet, P., 1986, Un système d'aide à la décision en prospection Uranifère: discrimination des anomalies géochimiques multiélémentaires significatives: Thèse de Doctorat d'Etat ès Sciences, Université de Nice, 436 p.

Poyet, P., 1990, Integrated access to information systems: Applied Artificial Intelligence, Hemisphere Publ. Co., v. 4, no. 3, p. 179-238.

Poyet, P., and Detay, M., 1988a, HYDROEXPERT: Aide à l'implantation d'ouvrages d'hydraulique villageoise: Proc. Eighth Int. Workshop on Expert Systems and their Applications, Avignon, France, v. 2, p. 397-410.

Poyet, P., and Detay, M., 1988b, L'avènement d'une génération de systèmes experts de terrain, *in* Demissie, M., and Stout, G. E., eds., Proc. of the Sahel Forum on the State-of-the-Art of Hydrology and Hydrogeology in the Arid and Semi Arid Areas of Africa, Ouagadougou, Burkina Faso, p. 567-577.

Poyet, P., and Detay, M., 1989a, Enjeux sociaux et industriels de l'intelligence artificielle en hydraulique villageoise: Proc. Première Convention Intelligence Artificielle, Hermes Publ., v. 2, Paris, p. 621-652.

Poyet, P., and Detay, M., 1989b, HYDROLAB: Un système expert de poche en hydraulique villageoise: Technique et Science Informatiques, v. 8, no. 2, p. 157-167.

Poyet, P., and Detay, M., 1989c, HYDROLAB: A new generation of compact expert systems: Computers & Geosciences, v. 15, no. 3, p. 255-267.

Poyet, P., and Detay, M., 1990, Artificial intelligence tools and techniques for water resources assessment in Africa, *in* Kuerzl, H., and Merriam, D.F., eds., Microcomputers applications in geology: Plenum Press Publ., New York, this volume.

Roche, C., and Laurent, J. P., 1989, Les approches objets et le langage LRO2 (KEOPS): Technique et Sciences Informatiques, v. 8, no. 1, p. 21-39.

Sandvad, E., 1989, Hypertext in an object-oriented programming environment: Proc. Workshop on Object Oriented Document Manipulation, France, BIGRE no. 63-64, p. 30-41.

Scheifler, R. W., and Gettys, J., 1986, The X window system: MIT Report of Athena Project, Cambridge, Massachussets, 29 p.

Sibertin-Blanc, C., 1988, Le modèle de données objet comme formalisme de modélisation d'une base de données: Modèles et Bases de Données, Juin 1988, no. 9, p. 27-40.

Sinding-Larsen, R., 1975, A computer method for dividing a regional geochemical survey area into homogeneous subareas prior to statistical interpretation, *in* Elliott, I. L., and Fletcher, W. K., eds., Geochemical exploration 1974: Elsevier Science Publ., Amsterdam, p. 191-217.

Sinding-Larsen, R., 1977, Comments of the statistical treatment of geochemical exploration data: Sci. Terre Inform. Geol., no. 9, p. 73-90.

Tryon, R. C. and Bailey, D. E., 1970, Cluster analysis: McGraw-Hill Book Co., New York, 347 p.

Tuckey, J. W., 1977, Exploratory data analysis: Addison-Wesley, Reading, Massachusetts, 506 p.

Tuckey, J. W., 1979, Robust techniques for the user, *in* Launer, R. L., and Wilkinson, G. N., eds., Robustness in statistics: Academic Press, New York, p. 103-106.

Valduriez, P., 1987, Objets complexes dans les systèmes de bases de données relationnels: Technique et Science Informatiques, v. 6, no. 5, p. 405-418.

Valenchon, F., 1982. The use of correspondance analysis in geochemistry: Jour. Math. Geology, v. 14, no. 4, p. 331-342.

ESTIMATING THE PROBABILITY OF OCCURRENCE OF MINERAL DEPOSITS FROM MULTIPLE MAP PATTERNS

Frederik P. Agterberg
Geological Survey of Canada, Ottawa, Ontario, Canada

ABSTRACT

By superimposing binary or other types of discrete map patterns on top of one another, a study area can be divided into many small map elements. In this paper, the probability of occurrence of mineral deposits is modeled as a function of the characteristics of such map elements. The two methods discussed for estimating this probability are weights of evidence modeling and logistic regression. Special consideration is given to the effects of selection of study area and map elements for a deposit type, intensity of exploration and spatial autocorrelation effects.

INTRODUCTION

Geoscience map patterns are used for decision-making in mineral exploration. Various prognostic indicators based on these patterns serve to predict future supplies from a region, or to rank target areas in order of priority before development consisting of more detailed surveys and drilling. Microcomputers equipped with geographic information systems are helpful for computation and display of prognostic indicators which generally are logical combinations of features known to be associated

Use of Microcomputers in Geology, Edited by D.F. Merriam
and H. Kürzl, Plenum Press, New York, 1992

with mineral deposits. Prognostic indicators can be expressed in terms of probabilities which are numbers between zero and one which obey the rules of probability calculus. In the statistical estimation of these probabilities, several factors must be considered: 1) Mineral deposits generated by different processes have different indicator variables which must be defined separately; 2) it is necessary to distinguish between probability of discovery and probability of existence; and, 3) although the map patterns are two-dimensional, it should be kept in mind that they are projections of three-dimensional objects which were formed over different periods of geological time.

Methods of multivariate analysis and expert systems already were used during the 1970s and early 1980s for estimating probabilities. Early prognostic maps (see e.g., Agterberg, 1984, with discussion by Tukey, 1984) were cell-based in that they displayed values estimated from variables quantified for cells belonging to grids superimposed on the map patterns. With the recent advent of geographic information systems, it has become practical to use, on microcomputers, numerous small, irregularly shaped homogeneous map elements which are separated by natural boundaries instead of by grid-lines. Such small map elements are used in weights of evidence modeling (Agterberg, Bonham-Carter, Wright, 1990; Bonham-Carter and Agterberg, 1990; Bonham-Carter, Agterberg, and Wright, 1990). Earlier methods (e.g., logistic regression analysis) also can be modified to compute probabilities for small map elements.

The similarity between problem-solving in mineral resource appraisal and logical reasoning in medical expert systems (Spiegelhalter, 1986; Lauritzen and Spiegelhalter, 1988) was discussed in Agterberg (1989a). Spatial analysis presents an additional requirement in mineral resource appraisal. This paper is organized as follows. The procedure for estimating the probability of occurrence of mineral deposits used in weights of evidence modeling is outlined first. This is followed by a detailed discussion of the methods required to handle spatial aspects and to use map elements instead of grid-cells.

BASIC RULES OF PROBABILITY CALCULUS

Before the advent of expert systems in the 1970s, there were two main methods for working with uncertainty: frequentist and Bayesian. In the frequentist approach, scientific measurements are modeled by using random variables, and the entire approach is underlain by the logically coherent framework of probability calculus. The Bayesians allow unmeasured, "subjective" prior probabilities as initial guesses which are modified on the basis of new evidence by using methods of mathematical statistics. The approach followed in this paper is "frequentist" in that the initial probabilities are measured and interpreted as the outcomes of stochastic experiments. Subjectivity is assumed to be restricted to definition of mineral deposits, indicator variables and study area.

The scientists who developed the first expert systems worked with subjective prior probabilities. Contrary to the Bayesians, they felt that the methods of mathematical statistics were too restrictive for the propagation of these probabilities. For this reason, systems of heuristic rules were developed. The two main systems at present are "fuzzy logic" (Zadeh, 1983; Duda, Gaschnig, and Hart, 1979) and Dempster-Schafer evidential reasoning (Schafer, 1987; Garvey, 1987; Chung and Moon, 1990). For a discussion of these systems and the use of probability theory, also see Kiiveri (1990).

The rules of heuristic systems can be regarded as approximations to the rules of probability calculus. For example, one rule of "fuzzy logic" (cf. Duda, Gaschnig, and Hart, 1979) is as follows. If n events all have a given probability of occurring, the probability that they will occur simultaneously is set equal to the smallest of the n probabilities. This rule generally gives better results than using the product of the probabilities of the n events. In evidential reasoning, different types of independent evidence in support of the same hypothesis are weighted relatively strongly, because the possibility of the hypothesis holding true may be ruled out if the supporting evidence is not logically consistent. Drawbacks of adopting a system with rules that are approximations to the rules of probability theory are: (1) approximations may be applied even when they are unnecessary (cf. Spiegelhalter, 1986); and (2) it is not possible to use methods of statistical inference to test the validity of heuristic rules. The basic rules for probabilities can be summarized as follows.

Suppose that $p(D|B)$ represents the conditional probability that event D occurs given event B (e.g., mineral deposit D occurs in a small unit cell underlain by rock type B). This conditional probability obeys three basic rules (cf. Lindley, 1987, p. 18):

(1) Convexity: $0 \leq p(D|B) \leq 1$ (D occurs with certainty if B logically implies D; then $p(D|B) = 1$, and $p(D^c|B) = 0$ where D^c represents the complement of D);

(2) Addition: $p(B \cup C|D) = p(B|D) + p(C|D) - p(B \cap C|D)$; and

(3) Multiplication: $p(B \cap C|D) = p(B|D) \cdot p(C|B \cap D)$.

These three basic rules lead to many other rules. For example, replacement of B by $B \cap D$ in the multiplication rule gives:

$$p(B \cap C \cap D) = p(B \cap D) \cdot p(C \mid B \cap D)$$

Likewise, it is readily derived that:

$$p(B \cap C \cap D) = p(C \mid B \cap D) \cdot p(B \mid D) \cdot p(D)$$

This leads to Bayes' theorem in odds form:

$$\frac{p(D \mid B \cap C)}{p(D^c \mid B \cap C)} = \frac{p(B \mid C \cap D)}{p(B \mid C \cap D^c)} \frac{p(D \mid C)}{p(D^c \mid C)}$$

or

$$O(D \mid B \cap C) = \exp(W_{B \cap C}) \cdot O(D \mid C)$$

where $O = p/(1-p)$ are the odds corresponding to $p = O/(1+O)$, and $W_{B \cap C}$ is the "weight of evidence" for occurrence of the event D given B and C. Suppose that the probability p refers to occurrence of a mineral deposit D within a small area on the map (unit cell). Suppose further that B represents a binary indicator pattern, and that C is the study area within which D and B have been determined. Theoretically, C is selected from an infinitely large universe (parent population) with constant probabilities for the relationship between D and B. In most applications only one study area is selected per problem and C can be deleted from the preceding expression. Then Bayes' theorem can be written in the form:

$$\ln O(D|B) = W_B^+ + \ln O(D); \;\; \ln O(D|B^c) = W_B^- + \ln O(D)$$

for presence and absence of B, respectively. If the area of the unit cell underlain by B is small in comparison with the total area of B, the odds O are approximately equal to the probability p.

As an example of this type of application of Bayes' theorem, suppose that a study area C, which is a million times as large as the unit cell, contains 100 deposits; 10 percent of C is underlain by rock type B, which contains 30 deposits. The prior probability $p(D)$ then is equal to 0.0001; the posterior probability for a unit cell on B is equal to 0.0003. The weights of evidence are $W_B^+ = 1.10$ and $W_B^- = -0.25$, respectively.

WEIGHTS OF EVIDENCE MODELING

The preceding formulation of Bayes' theorem can be used to estimate weights W_K ($K = A, B, C, \ldots$) for different geoscience features A, B, C, ... In weights of evidence modeling, it is attempted to use features which are conditionally independent of the point pattern of mineral deposits (D). In the situation of three map patterns, the probabilities satisfy:

$$p(A \cap B \cap C \cap D) = p(A \mid B \cap C \cap D) \cdot p(B \mid C \cap D) \cdot p(C \mid D) \cdot p(D)$$

If the map patterns are binary, any symbol for presence of a feature (e.g., A) can be replaced by its complement (e.g., A^c) indicating absence. In total, there are 16 probabilities which add to one. This system, therefore, has 15 degrees of freedom. It is noted that the order of the map patterns and the pattern of deposits can be interchanged in the preceding equation. It is convenient to keep D in the last position. If there are more than two states for some of the patterns, or if there are more than four patterns, the number of degrees of freedom is considerably larger than 15. In general, it is not possible to estimate all probabilities in the system directly from the patterns unless the number of deposits is very large. In a larger system, it is likely that many of the estimated probabilities would become zero and that the others would have large variances. In weights of evidence modeling, the preceding equation would be approximated by:

$$p(A \cap B \cap C \cap D) = p(A \mid D) \cdot p(B \mid D) \cdot p(C \mid D) \cdot p(D)$$

The number of degrees of freedom then is reduced to six, and all 16 probabilities can be calculated from six weights of evidence, for presence and absence of each of the three binary patterns, respectively. For example, if A is absent and B and C are present:

$$\ln O(A^c \cap B \cap C \cap D) = W_A^- + W_B^+ + W_C^+ + \ln O(D)$$

Each pair of weights (W^+, W^-) for presence or absence of a binary pattern is estimated from all mineral deposits in the study region. Consequently, the weights and the posterior probabilities are relatively precise. However, bias is introduced if the assumption of conditional independence is not satisfied. After estimation of all posterior probabilities by means of the weights of evidence method, it is possible to apply a chi-square or a Kolmogorov-Smirnov test for comparison of observed and expected frequencies of deposits by using the map of the posterior probabilities. A second type of significance test consists of evaluating the hypothesis of conditional independence (and other models) on all possible combinations of two or three map patterns (see next section for more detailed discussion).

The hypothetical example graphically represented in Figure 1 can be used to illustrate the strategy followed in weights of evidence modeling. In addition to D, there are nine map patterns in this example. The connecting lines in the graph represent relationships between patterns that are statistically significant. With the exception of E, all map patterns are related to D. The following equation expresses these relationships:

$$p(A \cap B \cap ... \cap J) = p(A \mid B \cap D) \cdot p(B \mid D) \cdot p(C \mid D \cap E) \cdot p(D) \cdot p(F \mid D) \\ \times p(G \mid D \cap H \cap I \cap J) \cdot p(H \mid D \cap I \cap J) \cdot p(I \mid D \cap J) \cdot p(J \mid D)$$

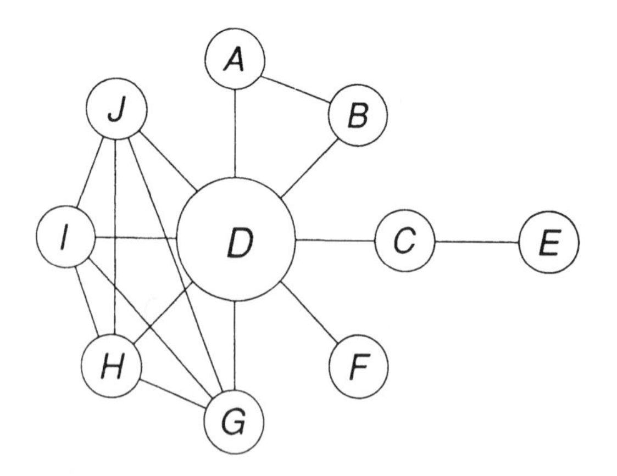

Figure 1. Hypothetical example of weights of evidence modeling. Mineral-deposit pattern D in center is related to nine map patterns A, B, ..., J. Relations between patterns are shown as solid lines. In weights of evidence modeling, posterior probabilities should be computed from weights for patterns that are conditionally independent of D. In order to achieve this, pattern E can be omitted because it is related only indirectly to D, and clusters of interrelated patterns (A, B and G, H, I, J) can be replaced by new patterns P_{AB} and P_{GHIJ}.

As it stands, only map pattern F is conditionally independent of the other patterns. The map patterns $G, H, I,$ and J are mutually interrelated in addition to being related to D. For example, they could represent a group of chemical elements that are all related to mineralization. In a situation of this type, addition of weights in the modeling is likely to create significant bias. In order to avoid this, one, for example, might perform first logistic regression analysis to replace the four patterns by a single probability index $\boldsymbol{P}_{\mathrm{GHIJ}}$. Then only two instead of eight separate weights would be estimated and the four patterns would not reinforce one another.

If the map patterns A and B would be conditionally independent of D, $p(A \mid B \cap D)$ simply would be replaced by $p(A \mid B)$. This substitution is not allowed, because A and B are related in the example. They could be combined into a single pattern according to one of the following two rules: $\boldsymbol{P}_{\mathrm{AB}} = p(A \cap B \mid D)$ if the measured joint probability for presence of the three events exceeds the value estimated by assuming conditional independence; or $\boldsymbol{P}_{\mathrm{AB}} = p(A \cup B)$ if it is less. After deleting E because it does not contribute directly information towards D, the preceding equation for weights of evidence modeling becomes:

$$p(A \cap B \cap \ldots \cap J) = \boldsymbol{P}_{\mathrm{AB}} \cdot p(C \mid D) \cdot p(F \mid D) \cdot \boldsymbol{P}_{\mathrm{GHIJ}}$$

The four remaining patterns could again be tested for approximate conditional independence before application. This type of test will be discussed briefly in the next section.

TESTING THE STRENGTH OF RELATIONSHIPS BETWEEN MAP PATTERNS

Methods of discrete multivariate analysis (see e.g. Bishop, Fienberg, and Holland, 1975) can be used for testing the assumption of conditional independence. Six models proposed in Agterberg (1990) are illustrated schematically in Figure 2. Models 1 to 4 are for two map layers (B and C) in addition to the point pattern of deposits (D), and models 5 and 6 are for three map layers (A, B and C) in addition to D.

Models 1 and 6 test for conditional independence by assuming that the relationships between the map layers are not significant statistically. If Model 2 is acceptable, C can be deleted because it independent

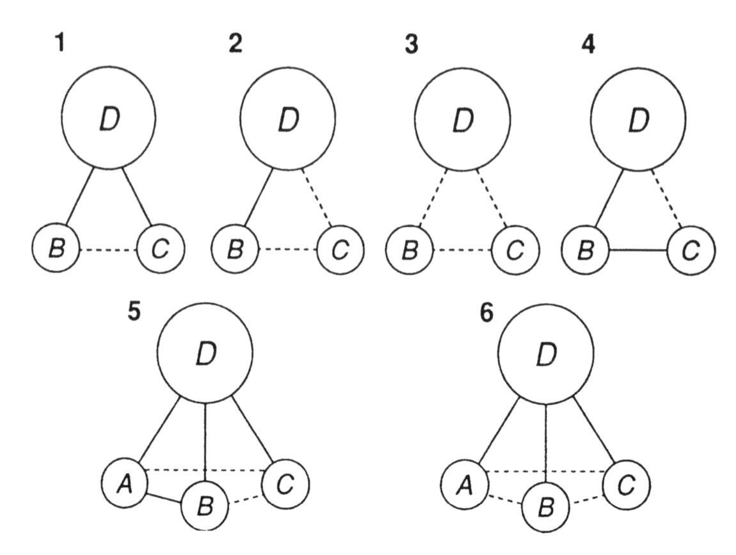

Figure 2. Six models (1 to 6) for testing strength of relationships between point pattern of mineral deposits (*D*) and sets of two or three map layers (*A*, *B*, and *C*). Solid lines are for statistically significant relations between patterns, and broken lines for relations that are not significant. See text for further explanation.

of *B* and *D*. Model 3 assumes that both *B* and *C* are independent of *D*. The usefulness of models 2 and 3 is limited because it is unlikely that map patterns that are not related to mineralization would be considered in practice. In model 4, *B* is related directly to *D* whereas *C* is related indirectly to *D*. In practice, this indicates that *C* should be used only if the information on *B* is missing. If *B* is known, *C* is redundant. Any results obtained from models 5 and 6 should corroborate those of model 1. In larger systems where all possible combinations of three map patterns are tested, models 5 and 6 are helpful in outlining groups of map patterns that are mutually interrelated such as *G*, *H*, *I*, and *J* in Figure 1.

The models of Figure 2 can be evaluated for goodness-of-fit by two types of chi-square test (see Bishop, Fienberg, and Holland, 1975). A model should not be used if the estimated chi-square value is too large provided that the spatial autocorrelation effects discussed at the end of this section are considered. Differences between successive models also can be tested by a chi-square test (steps from model 1 to 2, from 2 to 3, and from 5 to 6). The partial test for the step from model 5 to 6 is helpful in determining the extent to which the assumption of conditional independence is violated for specific pairs of map patterns. Contrary to models 1, 3, and 6, models 2, 4, and 5 yield different results when *B* and *C* are interchanged. This provides information on which one of a pair of map patterns is irrelevant (model 2), redundant (model 4), or conditionally independent with respect to a pair of other map patterns (model 5).

The chi-square tests for goodness-of-fit of the six models should be applied with caution. The test statistics are distributed asymptotically as chi-square with degrees of freedom determined by relationships between variables if the observations can be regarded as independent random trials. If the variables are map patterns, there are significant spatial autocorrelation effects which become large if the unit cell is decreased in size. Each test statistic can be divided into two parts, for comparing frequencies of unit cells with and without deposits, respectively. Only the first part (for unit cells with deposits) is independent of unit cell size and distributed as chi-square with one-half as many degrees of freedom as when all frequencies are used. Although they are not distributed approximately as chi-square, test statistics based on all frequencies (unit cells with and without deposits) remain useful for comparison with one another.

Problems of estimating the probability of occurrence of mineral deposits from multiple map patterns will be discussed by using the artificial examples of Figures 3 to 5.

WEIGHTS OF EVIDENCE; INTENSITY OF EXPLORATION; SPATIAL ANALYSIS

The effect of intensity of exploration is shown in Figure 3A-C where the study area is underlain by three rock types (*A*, *B*, and *C*). Part of the study area is not exposed because of a lake. Clearly, the number of discovered deposits increases in the course of time as a nondecreasing function of cumulative amount of exploration. Weights of evidence can be estimated only with relative precision if a sufficiently large number of deposits have been discovered. If n_C represents number of deposits underlain by map pattern *C*, then the variance of the weight for presence of *C* approximately satisfies $s^2(W_C^+) = 1/n_C$ (cf. Agterberg, Bonham-Carter, and Wright, 1990). Suppose that n_T represents total number of deposits in the test area (*T*); and that M_C and M_T are the areas of *C* and *T*, respectively. The unit cell area can be assumed to be small in comparison with M_C so that n_C is negligibly small in comparison with the total number of unit cells underlain by *C*. Then the weight of evidence for presence of *C* satisfies:

$$\exp(W_C^+) = p(C|D)/p(C|D^c) = (n_C/n_T)/(M_C/M_T)$$

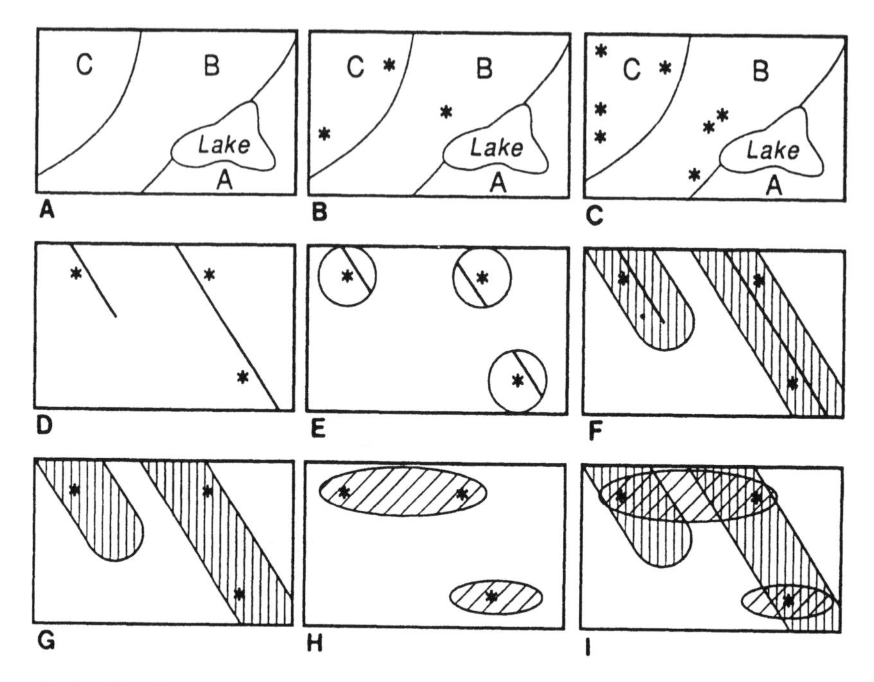

Figure 3. Artificial examples to illustrate effect of intensity of exploration (3A, B, C), spatial analysis (3D, E, F) and combining map patterns (3G, H, I). Prior probability is proportional to number of known deposits (*) in study area and increases as nondecreasing function of cumulative exploration; weights of evidence cannot be estimated if there are no known deposits as in 3A and for rock type A in 3B, 3C; on average, weights of evidence are independent of intensity of exploration; posterior probabilities for rock types B and C are proportional to number of known deposits per rock type in 3B, 3C. Deposits and lineaments in 3D were dilated by using circle in 3E and 3F, respectively. Total length of lineament segments within dilated pattern of 3E is related to number of deposits on corridors of 3F (see text for equation). Initially, prior probability is same for 3G and 3H; when two patterns are combined in weights of evidence modeling, posterior probability for an area of overlap 3F is obtained by applying weight of evidence of one pattern 3G or 3H to new prior probability that is set equal to posterior probability of other pattern 3H or 3G.

This is the likelihood ratio by which the prior odds $O(D) = n_T/M_T$ should be multiplied in order to obtain the posterior odds $O(D \mid C)$ for occurrence of a deposit in a unit cell. Because unit cell size is very small, the posterior probability is equal approximately to these odds, or $p(D \mid C) = n_C/M_C$. This first example illustrates the following three points:

(1) The weight of evidence for a map pattern depends on the ratio between frequencies of deposits on the map pattern and within the

total study area, respectively. If the intensity of exploration can be regarded as constant across the test area, this frequency ratio is constant and the estimate of the weight of evidence is unbiased and, on the average, independent of the intensity of exploration. If the intensity of exploration increases, the variance of the estimated weight of evidence decreases.

(2) The posterior probability for a unit cell on a map pattern is equal to the expected value of the random variable that can be defined for frequency of deposits per unit cell. This random variable, for example, would have a Poisson distribution if the deposits are distributed randomly across the map pattern.

(3) In general, the selection of study area may be arbitrary. For example, M_T would be changed if the lake in Figure 3A were to be excluded from T. Obviously, all measurements performed on a pattern are changed when the boundaries of the study area are moved. However, deposit density and weight of evidence for a map pattern are not biased by selection of study area, provided that the statistical relationship between the deposits and the map patterns does not change within the larger universe (parent population) from which the study area is selected.

The second example (Figure 3D-F) illustrates the spatial statistical analysis of map patterns (for deposit points D and lineaments L in this example) which is a field of research where considerable progress has been made during the past 10 years (cf. Berman and Diggle, 1989). Map patterns are regarded as sets.(A set consisting of line segments or plane figures contains an infinitely large number of points.) By using a geographic information system, measurements on dilated map patterns can be performed quickly. In dilation a map pattern is enlarged by superimposing a circle or other plane figure on all its points and including all additional points in the dilated pattern. For example, Figure 3E shows parts of lineaments contained in circles (with radius r) around the deposits; Figure 3F shows corridors which contain all points within a distance r from the lineaments. If the circle used for dilation is written as B_r, the dilated patterns can be written as $D \oplus B_r$ (Fig. 3E) and $L \oplus B_r$ (Fig. 3F), respectively.

Suppose that the combined length of the line segments in Figure 3E is written as $\mathbf{C}_1(r) = (D \oplus B_r) \cap L$; and the number of deposit points on the corridors of Figure 3F as $\mathbf{C}_2(r) = (L \oplus B_r) \cap D$. Two intensity measures, for average length of line and average number of deposit points per unit cell, can be denoted as $\wp_1 = M_L/M_T$ and $\wp_2 = n_T/M_T$. Then (cf., Stoyan and Stoyan, 1982), the spatial covariances $\mathbf{C}_1$ and $\mathbf{C}_2$ are related by the equation $\wp_1 \cdot \mathbf{C}_2(r) = \wp_2 \cdot \mathbf{C}_1(r)$. In general, $\wp_i \cdot \mathbf{C}_j(r) = \wp_j \cdot \mathbf{C}_i(r)$ (with $i \neq j$; $i,j = 1, 2$ or 3), because B_r also can be applied for dilation of areal patterns. Equations of this type are useful potentially because one type of spatial covariance $\mathbf{C}_i$ ($i = 1, 2$, or 3) can be computed from another. Stoyan and Stoyan (1982) have derived various measures of spatial influence of one pattern upon another from $\mathbf{C}_i$ (also see Stoyan and Ohser, 1982).

Corridors around linear features can be useful indicator patterns for mineral deposits. In Agterberg, Bonham-Carter, and Wright (1990), the following method was used to decide on the optimum width of the corridors. Weights W^+ and W^- are determined for increasing r; the contrast $C = W^+ - W^-$ is calculated for each successive corridor patterns; and the pattern with the largest value of C is selected. The contrast provides a measure of spatial correlation between an areal pattern and a point pattern. Its expected value is zero when the points are distributed randomly across the study area.

Figures 3G-I illustrate how two binary patterns are combined with one another in weights of evidence modeling. Suppose that the two patterns are termed B and C. In the area of overlap (see Fig. 3I), the prior probability is multiplied by $\exp(W_B^+ + W_C^+)$. The two patterns reinforce one another as follows. If B is known to be present, the prior probability is changed from n_T/M_T to n_B/M_B, before it is multiplied by the likelihood ratio for C which is equal to $\exp(W_C^+)$ if C also is present. Likewise, presence of C changes the prior probability for B.

THREE-DIMENSIONAL MODELING; LACK OF CONDITIONAL INDEPENDENCE; CLUSTERING

A potentially useful three-dimensional extension of two-dimensional dilation is shown in Figures 4A-C. Suppose that the pattern in Figure 4A is for two intrusive rock bodies with contacts that dip steeply to the north (= toward top of diagram) and gently to the south. Ordinary dilation leads to the pattern of Figure 4B but three-dimensional dilation perpendicular to the contacts would yield a different, probably more realistic pattern as illustrated in Figure 4C. The three-dimensional aspects of relations between mineral deposits and rock types become relatively more important when the scale of the geoscience map is increased.

Figures 4E-F show two possible violations of the assumption of conditional independence which is assumed to be satisfied in Figure 4D.

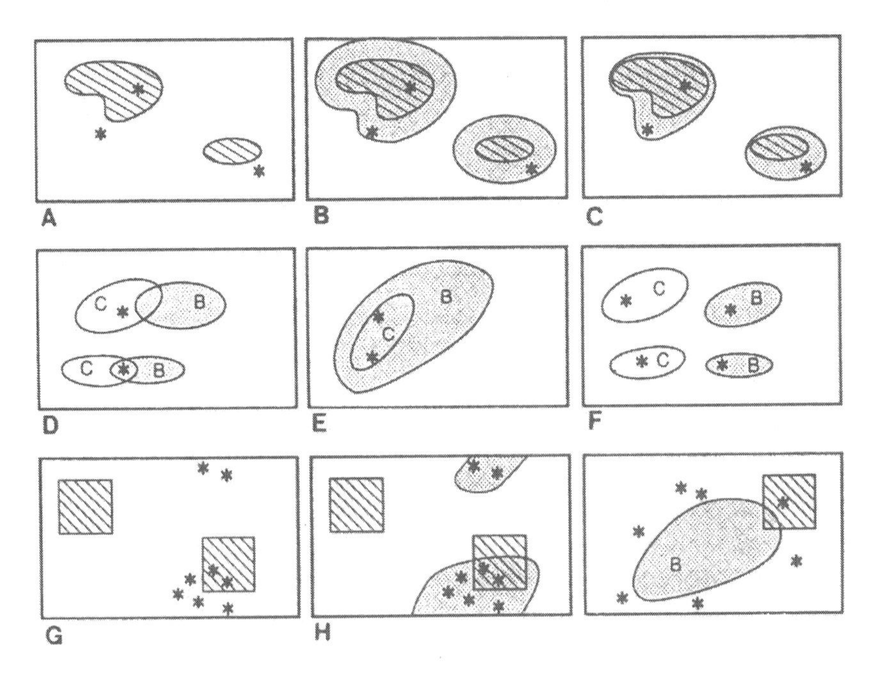

Figure 4. Artificial examples to illustrate three-dimensional effect of depth to contact of intrusive rock (4A, B, C), lack of conditional independence (4D, E, F) and clustering (4G, H, I). Amount of dilation is assumed to depend on strike and dip of contact in 4C. Deposits are restricted to rock type *C* which is contained within *B* in 4E; *B* and *C* are without overlap in 4F. This suggests that it may be possible to delete *B* from 4E, and combine *B* and *C* into single pattern in 4F. Clusters of mineral deposits without and with additional information are shown in 4G and 4H; deposits cluster around contact of rock type *B* in 4I. See text for further explanations.

Posterior probabilities for unit cells within the area of overlap of the two map patterns predicted by using the assumption of conditional independence are less and greater than those observed in Figures 4E and 4F, respectively. A possible strategy for avoiding biased posterior probabilities (also see previous discussion of Fig. 1) then is to combine *B* and *C* into a single new map pattern using $B \cap C$ (for Fig. 4E) and $B \cup C$ (for Figure 4F), respectively.

Problems of clustering of mineral deposits without and with additional information are graphically illustrated in Figures 4G-I. The probability that a cell *Q* (similar the squares in Fig. 4G) placed at random on the study area *T* contains one or more deposits can be measured by dilating the deposits *D* by *Q*. If *M* denotes area, this probability is equal to the ratio $M(D \oplus Q)/M(T)$. Suppose that the occurrence of *D* is restricted to a smaller area where a known indicator variable is present (see Fig. 4H). Then the number of deposits per cell Q is a discrete random variable controlled by relative area of the indicator variable in the cell. For example, if this relative area is a continuous random variable with the gamma distribution, then the number of deposits per cell satisfies a negative binomial distribution (cf. Agterberg, 1984).

In Figure 4I, the deposits are spatially related to the contact of rock type B. In order to characterize this type of situation, Agterberg and Fabbri (1978) measured the ratio $M\{B \cap (D \oplus Q)\}/M(D \oplus Q)$ and regarded it as an estimate of the mean of the random variable Y for the relative area of B in a cell Q with a deposit at its center. The random variable Y was related to random variable $X = M\{B \cap (P \oplus Q)\}/M(P \oplus Q)$ where P represents a point with random location within the study area. This resulted in an estimate of the relative area of B in cells most likely to contain a deposit. In the situation of Figure 4I, this optimum proportion value would be slightly less than 50 per cent because the deposits occur outside the curved contact of B.

CONTOUR MAPS; GRIDDING; MAP ELEMENTS AND MAP ELEMENT TYPES

Figure 5A is an example of a contour map on which the mineral deposits are associated with the anomaly with the highest values. In this situation, the contour map can be changed into a binary map with separate weights of evidence for presence and absence of the highest values. The threshold value (= 4.0 in Fig. 5A) can be optimized by estimating the contrast $C = W^{+} - W^{-}$ for successive contour values and selecting the contour with the largest contrast. This procedure cannot be followed in the situation of the contour map of Figure 5B which can be changed into a binary map by taking two threshold values. Figure 5C shows the combination of the binary pattern of Figure 5B with another binary pattern. A general method for estimating a variable weight function $W^{+}(x)$ which depends on contour value x has been proposed in Agterberg and Bonham-Carter (1990).

The problem of loss of precision when variables are quantified by using grid cells is illustrated in Figures 5D-F. In Figures 5D and 5E the presence of a feature in a cell is indicated by a plus sign at the cell center. Contour values at cell centers can be used in the situation of Figure 5F. An advantage of using relatively large grid cells for coding is that these define neighborhoods of coexisting variables. Thus new variables can be

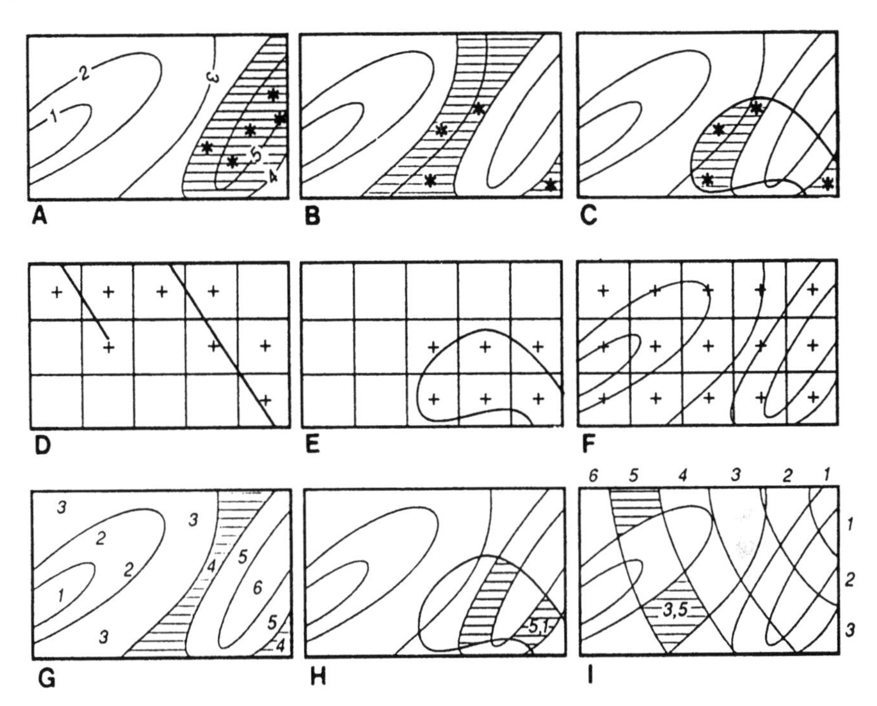

Figure 5. Artificial examples to illustrate reduction of contour maps to binary map patterns (5A, B, C), gridding (5D, E, F) and definition of map elements using contour maps (5G, H, I). Binary values between 2.5 and 4.0 respectively. Pattern of 5B is combined with pattern for rock type in 5C. Crosses in 5D and 5E are for presence of lineaments and rock type in grid cells. Contour values at cell centers can be used in 5F. Superposition of rock type pattern and another contour pattern on contour pattern of 5G are illustrated in 5H, and 5I, respectively. Different map elements belong to same types as follows: type 4 in 5G; (5,1) in 5H; and (3,5) in 5I. See text for further explanations.

defined for features which are in each other's vicinity. For example, many types of deposits tend to be associated with contacts between two rock types. Although a deposit may be underlain by one of these rock types on the geological map that it used, the presence of the other rock type in the immediate vicinity then significantly increases the probability of occurrence of a deposit. If regression analysis (cf. Agterberg, 1984) is used for estimation, weights are computed for all indicator variables. The use of product variables $x_{ij} = x_i \cdot x_j$, where x_i and x_j represent relative cell area underlain by two rock types i and j, then allows more weight to be assigned to coexistence of two rock types in the same cells. In general, one or more of the product variables x_{ij} provide better indicator variables than x_i and x_j taken separately.

Nevertheless, it is likely that any advantage of using grid cells for characterizing neighborhoods is offset by the drawbacks resulting from loss of precision. In general, it is better to work with small, homogeneous map elements as illustrated previously for binary patterns (e.g., Figs. 3A and 3I), and for contour maps in Figures 5G-I. Spatial associations such as contacts between rock types then can be modeled by morphological operations such as dilation. When p indicator variables are defined for a study area, every map element has p values for these variables. Nonadjacent map elements can have the same set of values for the indicator variables as illustrated in Figure 5H where two map elements of map element type (5,1) have value 5 for the variable based on the contour map of Figure 5G, and value 1 for presence of the rock type. The example of Figure 5I shows two map elements of type (3,5) with values derived from two contour maps.

WEIGHTED LOGISTIC REGRESSION ANALYSIS

The probability p_i that a mineral deposit occurs in a small unit cell located in a map element belonging to type i ($i = 1,2,\ldots,n$) can be estimated by logistic regression with each map element type weighted according to: (1) the number of deposits it contains, and (2) the combined area of its map elements. This procedure will be described here. It is followed in the FORTRAN 77 program LOGPOL which is a modified version of the microcomputer program LOGDIA (Agterberg, 1989b). Applications of LOGPOL are described in Agterberg (1990) and Reddy, Agterberg, and Bonham-Carter (in press).

In logistic regression, the probability p_i is derived from its logit $v_i = \ln \{p_i/(1-p_i)\} = X_i\alpha$ where X_i is the row vector of values for the indicator variables and α is a column vector of unknown coefficients which are to be estimated. Suppose that m_i represents the number of deposits in map element type i, and that m_i is greater than zero for n_d map element types. It is convenient to relabel m_i as m_j ($j = 1,2,\ldots,n_d$) after deleting all values $m_i = 0$. The total number of unit cells belonging to map element type i can be written as t_i. In general, t_i is not an integer number because it represents total area of the map element type with the unit cell used as unit of area. If the size of the unit cells is sufficiently small, the number of unit cells without deposits per map element type i can be set equal to t_i ($i = 1,2,\ldots,n$).

The input data block for logistic regression can be divided into two parts, for unit cells in map element types with and without deposits, respectively. A column vector Y with $N = n_d + n$ elements can be defined to distinguish between these two parts. The top part of Y consists of n_d ones for map element types with deposits only, and its bottom part has n zeros for all map element types. Consequently, $Y_j = 1$ if $j \leq n_d$, and $Y_j = 0$ if $n_d < j \leq N$.

The j-th row ($j = 1,2,\ldots,N$) of the data block for logistic regression consists of the j-th row of a column matrix X with values for the indicator variables, an element of the column vector Y, and a weight w_j. The first n_d rows of X are for relabeled map element types with one or more deposits. In the data block, they have $Y_j = 1$, and weights $w_j = m_j$ ($j = 1,2,\ldots,n_d$). The remaining n rows of X are identical to the previously defined rows X_i ($i = 1,2,\ldots,n$) with $Y_j = 0$ ($j = n_d+1,\ldots,N$), and weights $t_j \equiv t_i$. Suppose now that V is an ($N \times N$) diagonal matrix with nonzero elements $V_{jj} = w_j \cdot Y_j \cdot (1-Y_j)$ where Y_j is an estimated value of the probability p_j. (It is noted that $p_j \equiv p_i$ for $i = j - n_d = 1,2,\ldots,n$.) If the maximum likelihood method is used for estimation, a column vector of scores $S = Y - Y$ is made to converge until the relation $X'S = 0$ is satisfied.

Suppose that the subscript k is used to distinguish between successive estimates during the iterative process preceding the final result. Before convergence is reached, successive approximations $\alpha_1, \alpha_2, \ldots, \alpha_k, \ldots$ of the vector of coefficients (α) satisfy:

$$\alpha_k = (X'V_{k-1}X)^{-1}X'V_{k-1}(X\alpha_{k-1} + V^{-1}_{k-1}S_{k-1})$$

At the beginning of the iterative process, a preliminary estimate α_0 is used to obtain S_0 and V_0, leading to α_1. Setting all coefficients of α_0 equal to zero normally leads to satisfactory results in fewer than ten iterations. After convergence, the (final) estimated coefficients α are used to calculate the logits ν_i ($i = 1,2,\ldots,n$) and this results in the (final) estimated probabilities.

In ordinary logistic regression, the goodness-of-fit of the model is evaluated by using a chi-square test and the deviance statistic. In the preceeding weighted version for map element data, these two test statistics depend on unit cell size. If the unit cells are sufficiently small, a further decrease in their size does not result in changes in the estimates of the coefficients except for the constant term which continues to decrease. (Each decrease in the constant term is equal to the natural logarithm of the factor by which the unit cell size is decreased.) For smaller unit cell sizes, the chi-square and deviance statistics continue to increase. (The chi-square increases more rapidly than the deviance.) It is, however, possible to evaluate the goodness-of-fit of the model by classifying the deposits by using the probabilities estimated for the map element types to which they belong. Observed and expected frequencies of mineral deposits then are compared by means of the same type of test (chi-square or Kolmogorov-Smirnov test) used for the evaluation of posterior probability maps in weights of evidence modeling.

CONCLUDING REMARKS

An advantage of logistic regression with respect to weights of evidence modeling is that bias due to lack of conditional independence of the map patterns is avoided. On the other hand, weights of evidence modeling has the advantage of greater flexibility: each map pattern has two coefficients (W^+ and W^-) instead of one. The estimated weights of evidence of a map pattern and their variances are independent of those of the other map patterns and can be used when information on one or more of the other map patterns is not available at a particular place.

ACKNOWLEDGMENTS

Thanks are due to Graeme Bonham-Carter for critical reading of the manuscript and to Ramesh Reddy for helpful suggestions.

REFERENCES

Agterberg, F.P., 1984, Use of spatial analysis in mineral resource evaluation: Jour. Math. Geology, v. 16, no. 6, p. 565-589.

Agterberg, F.P., 1989a, Computer programs for mineral exploration: Science, v. 245, no. 4913, p.76-81.

Agterberg, F.P., 1989b, LOGDIA-FORTRAN 77 program for logistic regression with diagnostics: Computers & Geosciences, v. 15, no. 4, p. 599-614.

Agterberg, F.P., 1990, Combining indicator patterns for mineral resources evaluation: Preprint Proc., Intern. Workshop on Statistical Prediction of Mineral Resources, held at China Univ. of Geosciences, Wuhan, China, October 20-25, 1990, v. 1, p. 1-16.

Agterberg, F.P., and Bonham-Carter, G.F., 1990, Deriving weights of evidence from geoscience contour maps for the prediction of discrete events: Proc., 22nd APCOM Symposium, held in Berlin, Germany, September 17-21, 1990, Techn. Univ. Berlin, v. 2, p.381-396.

Agterberg, F.P., and Fabbri, A.G., 1978, Spatial correlation of stratigraphic units quantified from geological maps: Computers & Geosciences, v. 4., no. 3, p. 285-294.

Agterberg, F.P., Bonham-Carter, G.F., and Wright, D.F., 1990, Statistical pattern integration for mineral exploration, *in* Gaal, G., and Merriam, D.F.. eds., Computer applications in resource exploration: Pergamon Press, Oxford, p. 1-15.

Berman, M., and Diggle, P., 1989, Estimating weighted integrals of the second-order intensity of a spatial point process: Jour. Royal Stat. Soc. B, v. 51, no. 1, p. 81-92.

Bishop, M.M., Fienberg, S.E., and Holland, P.W., 1975, Discrete multivariate analysis: theory and practice: MIT Press, Cambridge, Massachusetts, 587p.

Bonham-Carter, G.F., and Agterberg, F.P., 1990, Application of a microcomputer-based Geographic Information System to mineral-potential mapping, *in* Hanley, J.T., and Merriam, D.F., eds., Microcomputer applications in geology II; Pergamon Press, Oxford, p. 49-74.

Bonham-Carter, G.F., Agterberg, F.P., and Wright, 1990, Weights of evidence modelling: a new approach to mapping mineral potential, *in* Statistical applications in the earth sciences: Geol. Survey Canada Paper 89-9, p. 171-183.

Chung, C.F., and Moon, W.M., 1990, Combination rules of spatial geoscience data for mineral exploration: Proc., Intern. Symp. on Mineral Exploration: The Use of Artificial Intelligence, held in Tokyo, October 29-31, and Tsukuba, Japan, November 1-2, 1990.

Duda, R.O., Gaschnig, J.G., and Hart, P., 1979, Model design in the PROSPECTOR consultation system for mineral exploration, *in* Mitchie, D., ed., Expert systems in the microelectronic age: Edinburgh Univ. Press, Edinburgh, p. 153-167.

Garvey, T.D., 1987, Evidential reasoning for geographic evaluation for helicopter route planning: IEEE Transactions on Geoscience and Remote Sensing, v. GE-25, no. 3, p.294-304.

Kiiveri, H.T., 1990, On the integration of spatially related databases and the representation and propagation of uncertainty in images: DMS Report WA 90/5, CSIRO, Wembley, Western Australia, 23p.

Lauritzen, S.L., and Spiegelhalter, D.J., 1988, Local computations with probabilities on graphical structures and their application to expert systems: Jour. Royal Statist. Soc. B, v. 50, no. 2, p. 157-224.

Lindley, D.V., 1987, The probability approach to the treatment of uncertainty in artificial intelligrance and expert systems: Statistical Science, v. 2, no. 1, p. 17-24.

Reddy, R.K.T., Agterberg, F.P., and Bonham-Carter, G.F., in press, Application of GIS-based logistic models to base-metal potential mapping in Snow Lake area, Manitoba: Proc., 3rd Intern. Conf. on Geographic Information Systems, Ottawa, Canada, March 18-21, 1991.

Schafer, G., 1987, Probability judgement in artificial intelligence and expert systems: Statistical Science, v. 2, no. 1, p. 3-16.

Spiegelhalter, D.J., 1986, Uncertainty in expert systems, *in* Gale, W.A., ed., Artificial intelligence and statistics: Addison-Wesley, Reading, Massachusetts, p. 17-55.

Stoyan, D., and Ohser, J., 1982, Correlations between planar random structures with an ecological application: Biom. Jour. Berlin, v. 24, no. 7, p.631-647.

Stoyan, D., and Stoyan, H., 1982, Quantifizierung von Korrelationen zwischen geometrischen Strukturen auf geologischen Karten: Zeitschrift fHr angewandte Geologie, v. 28, no. 5, p.240-244.

Tukey, J.W., 1984, Comments on Use of spatial analysis in mineral resource evaluation: Jour Math. Geology, v. 16, no. 6, p. 591-594.

Zadch, L.A., 1983, The role of fuzzy logic in the management of uncertainty in expert systems: Fuzzy Sets and Systems, v. 11, p. 199-227.

USE OF A LAPTOP COMPUTER AND SPREADSHEET SOFTWARE FOR GEOPHYSICAL SURVEYS

Robert S. Sternberg
Franklin and Marshall College
Lancaster, Pennsylvania, USA

ABSTRACT

The use of a laptop computer and spreadsheet software have been introduced to geophysical field work for data acquisition, logging, reduction, and plotting. This is a helpful method for many types of geophysical problems, and can support the geophysicist in the field to understand what is happening as the survey progresses. Examples are given of the application to seismic data acquisition, gravity data reduction, and gravity modeling.

INTRODUCTION

Geophysical survey equipment usually involves rapid data collection. This is especially true for surveys covering a small area where the operators can walk between stations, and where terrain is not an obstacle. For such surveys, the traditional jotting down of readings in a field notebook, along with later entering of data into an office computer, may not be the most efficient way to proceed. Seismic traces, although not collected rapidly, consist of many data points. For other data types, such as gravity, the readings must be reduced to account for various corrections, so the meaning of the raw readings may not be readily

Use of Microcomputers in Geology, Edited by D.F. Merriam
and H. Kürzl, Plenum Press, New York, 1992

apparent. In addition, even simple modeling of corrected data in the field can yield a better understanding of the survey as it develops. Hence, digital data acquisition and data reduction, modeling and graphing in the field may save time in the long run, can give an immediate feel for what is going on with the survey, and may even be useful interactively to guide the survey in new directions.

The availability of compact and relatively inexpensive laptop computers, which can run in the field off rechargeable batteries, and generic spreadsheet software provides one approach for working with the data while in the field. Jones (1983) and Kelly, Dale, and Haigh (1984), for example, discuss the use of microcomputers for various types of geophysical field work. Tarkoy (1986) describes generic software, including, spreadsheets, useful for applications in engineering geology. These tools are useful particularly for teaching exploration geophysics to undergraduate students taking their first or second geophysics course.

COMPUTER HARDWARE

Laptop computers are useful not only for the traveling business executive seen in many advertisements, but also for the traveling scientist, including the field geophysicist. We have used an IBM-compatible laptop computer (NEC Multispeed) as a field computer. The computer is a typical laptop: it runs on wall current or for about 4 hours from rechargeable batteries, has 640K memory, two 3-1/2" 720K disk drives, a 25-line liquid crystal display graphics monitor, weighs 5 kg, and costs under $1500. Hard disks, improved backlit screens, and modems also are available for laptops. We also have an inkjet printer (Diconix 150) which can run from wall current or rechargeable batteries, weighs 1.7 kg, and costs about $400. The computer and printer together fit into a briefcase-sized bag. Drawbacks to this equipment are that a single computer battery pack is not adequate for a full day of field work, and this equipment is not intended for use in rugged conditions (dirt, temperature, humidity, etc.). The former problem, at least, can be circumvented with the use of a supplementary battery pack, which can be purchased for about $100.

Some geophysical instruments also are able to store data digitally. This is a feature available, for example, in several models of proton precession magnetometers. The magnetometer data then can be uploaded to the computer while in the field. It is possible also to construct digital interfaces for certain instruments. As discussed briefly here (and also in Spadafore and Sternberg, 1990), we have constructed such an interface for our one-channel seismograph.

SPREADSHEET SOFTWARE

Spreadsheets, modeled after the accountant's ledger, also are useful for scientific and pedagogic applications (e.g., Herweijer, 1986; Tarkoy, 1986; Dory, 1988; Guglielmino, 1989; Walter, 1989). Either data or text can be entered into columns, rows, or the individual cells where a particular row and column intersect. Numerical operations can be performed on the data using equations entered into the cells; these can incorporate constants embedded in the spreadsheet as well as intrinsic spreadsheet mathematical functions. Equations can be copied quickly to a group of related cells. For example, column 1 might contain values of an independent variable x. The cell in row 1 of column 2 could contain the equation representing the dependent variable y which depends on the value of x in row 1 of column 1 and other constants. This equation can be quickly copied to the other rows of column 2, with each calculated value of y using the corresponding value of x. If the constants or independent variables are changed, the dependent variables expressed by these equations will be updated automatically.

Using a spreadsheet has some similarity to computer programming, but spreadsheet templates, as spreadsheet programs usually are termed, can offer several advantages over conventional programming. Constructing a spreadsheet may be simpler than writing a computer program; the basics needed to understand and construct spreadsheet templates can be learned in a few hours. The spreadsheet visually emphasizes the interrelationships between the different parts of a problem. The spreadsheet allows models to be gradually constructed. Finally, the spreadsheet easily permits data, model parameters, or model structures to be altered in a "what-if" fashion. Graphics and some database capabilities are included in many spreadsheet packages, which usually can run from one or two 720K floppy disks, 256K of computer memory, and cost about $200. We use the spreadsheet Quattro in the field, although the printouts included here were generated by transferring the templates (with minor modifications) to the spreadsheet program *Excel on a Macintosh* computer, and using a laser printer in the laboratory.

APPLICATIONS

Seismics - An Example of Data Acquisition

For seismic refraction surveys we have been using a single-channel signal-enhancement seismograph (Bison Model 1570C). In our earlier work, first breaks were picked off the screen in the field, after which the traces were erased. However, permanent records of the traces are desirable in order to compare adjacent traces, check the picks for first breaks, and possibly to detect additional phases. Bison does offer a stripchart recorder and a digital cassette tape recorder to which traces can be dumped. The former gives only an analog record, and the latter was a bit expensive for our use.

In order to obtain digital records that could be examined in the field, we took a different approach. Steve Spadafore of Franklin and Marshall College designed and built a simple circuit to convert the seismograph communications port from parallel to serial. Schematics of this circuit will be given in Spadafore and Sternberg (1990), or can be obtained from the author. The 256 points comprising each trace are uploaded at 9600 baud into the computer and saved to disc in a text file using a BASIC program (also available from author). The data in the file can be loaded into a spreadsheet template and plotted from the spreadsheet (Fig. 1). State-of-the-art multichannel seismographs already have a similar computer interface as part of the instrument. The power of our seismograph has been amplified considerably with this addition to our system.

Gravity - An Example of Data Reduction

Spreadsheet capabilities are well illustrated by an example from gravity surveying. Collection of gravity data is relatively slow, so the data logging is not a limiting factor. Nonetheless, the laptop/spreadsheet can serve as an electronic field book for logging the raw data. In addition, the spreadsheet can be designed to make the necessary gravity corrections, so gravity anomalies can be calculated and plotted while in the field. Jones (1983) has done these reductions using an HP41 calculator. The ledger-like nature of the spreadsheet makes it suitable to apply repeatedly the same gravity correction formulae to measurements made at different stations. Herweijer (1986) used a spreadsheet approach similar to the one described here for reduction of gravity data once back in the office, but doing this while in the field is instructive as the survey develops. Sutter and Sternberg (1988) also used a spreadsheet for

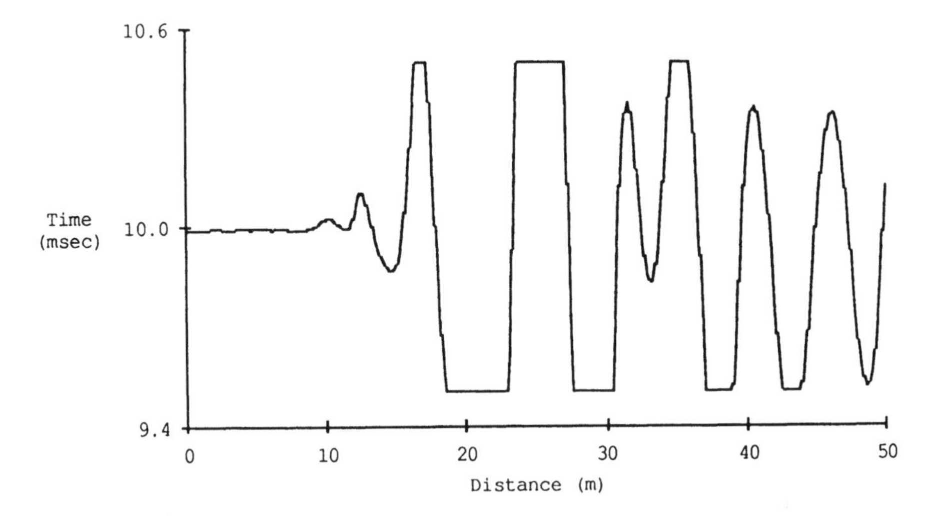

Figure 1. Seismograph trace uploaded into field computer and plotted from spreadsheet template. Trace consists of 256 points equally spaced in time. Units of relative amplitude are offset for shot-geophone distance of 10 m.

tabulation of gravity survey data, although the data reduction was done actually with a linked BASIC program. Their approach, although somewhat more difficult than the straight spreadsheet approach advocated here, can become advantageous for larger projects or for more complicated types of data reduction (e.g., topographic and tidal).

A spreadsheet template for the reduction of gravity data is shown in Figure 2, displaying the numerical results as the spreadsheet generally appears to the user. Header information describes the program and the survey for which it is being used. Inputs include: the factors needed to calibrate from dial units to mgal on the LaCoste and Romberg gravimeter (cells B44-B46, where the letter designates the column and the number designates the row); initial and final base station readings used for the drift correction (B50-C51); latitude (B54) and the Bouguer gravity anomaly (B57) of the base station used for the latitude correction and calculation of anomalies relative to the base station; Bouguer density used for the Bouguer correction (B60). Run constants include: the drift rate determined from the base station readings (cell F49); conversion factors from feet-to-meters and from mgal-to-gravity units (F54, F55); the free-air correction factor (F56); the Bouguer correction factor calculated for the given Bouguer density (F57); and the latitude correction factor calculated for the given base station latitude (F58). For each

	A	B	C	D	E	F	G	H	I
1									
2			GRAVITY SURVEY DATA REDUCTION						
3									
4	**								
5									
6	Name:	Grav-Red							
7									
8	Author:	Rob Sternberg							
9		Franklin & Marshall College							
10	Date:	Apr-88							
11		Slightly modified for Excel: Dec-89							
12									
13	Purpose:	To reduce LaCoste and Romberg gravity survey data to							
14		corresponding gravity anomalies.							
15		Default graph is total anomaly vs. N-S distance.							
16									
17	Instructions:								
18		1 Enter any notes concerning survey.							
19		2 Enter gravimeter dial units-to-mgal calibration data.							
20		Use the range that is most applicable.							
21		3 Enter base station readings and times.							
22		Assumes one inital and one final base station reading.							
23		Second reading may be entered after it is taken.							
24		4 Enter latitude of base station for latitudinal correction.							
25		5 Enter absolute gravity anomaly at base station.							
26		6 Enter density for Bouguer slab.							
27		7 Enter data for survey stations.							
28									
29	References:		LaCoste and Romberg Instruction Manual						
30									
31			An Introduction to Geophysical Prospecting						
32			P. Kearey and M. Brooks						
33			Blackwell, Oxford, 1984, ch. 6						
34									
35	**								
36									
37	SURVEY NOTES								
38	Date	Mar-88		Crew	Geo 37 class				
39	Location	southern Lancaster county							
40	Purpose	To investigate the south-north gravity gradient							
41	Notes	Latitude correction relative to 40.03 deg N							
42									
43	GRAVIMETER CALIBRATION INPUTS								
44	dial	3700							
45	mgal	3785.98							
46	factor	1.02408							
47									
48	BASE STATION INPUTS				BASE STATION DRIFT CALCULATIONS				
49	base	dial	time(min)		drift	-0.001	dial/min		
50	1	3702.044	41						
51	2	3701.959	210						
52									
53	BASE STATION LATITUDE INPUT				CONSTANTS				
54	latitude	40.03	degrees		ft-to-m	0.3048	m/ft		
55					mgal-gu	10	gu/mgal		
56	BASE STATION ANOMALY INPUT				free air	3.086	gu/m		
57	delta g	-558.90	gu		Bouguer	1.13E+00	gu/m		
58					latitude	7.998	gu/km		
59	BOUGUER DENSITY INPUT								
60	density	2700	kg/m3						
61									

Figure 2. Spreadsheet template for reduction of gravity data.

62									
63				STATION INPUTS	(DATA)				
64	**								
65	Station	N-S dist.	elev.	time	reading				
66		km	ft	min.	dial				
67	1	-13.60	400.0	89	3712.388				
68	2	-11.70	260.0	107	3719.501				
69	3	-9.70	340.0	139	3713.413				
70	4	-7.60	420.0	160	3706.677				
71	5	-4.70	260.0	156	3713.029				
72	6	-2.95	240.0	196	3711.941				
73	7	-0.70	360.0	212	3704.315				
74	base 1	0.00	400.0	41	3702.044				
75									
76				STATION	CALCULATIONS				
77	**								
78	elev.	drift	calibrate	convert	latitude	free air	Bouguer	from base	total
79	m	dial	mgal	gu	gu	gu	gu	gu	gu
80	121.9	3712.412	3798.691	37986.91	38095.68	38471.93	38333.97	214.95	-343.95
81	79.2	3719.534	3805.985	38059.85	38153.42	38397.98	38308.31	189.29	-369.61
82	103.6	3713.462	3799.766	37997.66	38075.25	38395.05	38277.79	158.77	-400.13
83	128.0	3706.737	3792.879	37928.79	37989.58	38384.63	38239.77	120.76	-438.14
84	79.2	3713.087	3799.382	37993.82	38031.41	38275.97	38186.30	67.28	-491.62
85	73.2	3712.019	3798.288	37982.88	38006.48	38232.23	38149.45	30.43	-528.47
86	109.7	3704.401	3790.487	37904.87	37910.47	38249.09	38124.92	5.91	-552.99

Figure 2. Continued

station, inputs (cells A67-E74) include: station number; distance of the station along the traverse; elevation of the station used for the elevation correction; time of the reading used for the drift correction; and the reading itself. Finally, the spreadsheet then calculates for each station (rows 80-86, in successive columns A-I): the elevation in meters; the drift-corrected dial reading; the calibrated mgal readings; the equivalent value in gravity units; the latitude-corrected result; the free-air anomaly; the simple Bouguer anomaly; the Bouguer anomaly relative to the base station; and the absolute Bouguer anomaly.

The same spreadsheet is shown in Figure 3 using the option to display the formulae that were entered originally into the template. The text in the upper part of the spreadsheet is not shown, and some of the column headers are truncated. The equation for the latitude correction in cell F58 illustrates how equations are entered. This equation is given algebraically as

$$\text{latitude correction} = 8.12 \sin(2\varnothing) \text{ gravity units/kilometer,} \qquad (1)$$

where $\varnothing$ is the latitude of the base station. The spreadsheet recognizes the entry in this cell as an equation rather than a number or character

	A	B	C	D	E	F	G	H	I
43	GRAVIMETER								
44	dial	3700							
45	mgal	3785.98							
46	factor	1.02408							
47									
48	BASE STATIO				BASE STATION DRIFT (				
49	base	dial	time(min)		drift	=(B2 -B1)/(TB2-TB1)	dial/min		
50	1	3702.044	41						
51	2	3701.959	210						
52									
53	BASE STATIO				CONSTANTS				
54	latitude	40.03	degrees		ft-to-m	0.3048	m/ft		
55					mgal-gu	10	gu/mgal		
56	BASE STATIO				free air	3.086	gu/m		
57	delta g	-558.9	gu		Bouguer	=0.0004191*RHO	gu/m		
58					latitude	=8.12*SIN(2*LAT*PI()/180)	gu/km		
59	BOUGUER DEN								
60	density	2700	kg/m3						
61									
62									
63									
64	***********								
65	Station	N-S dist.	elev.	time	reading				
66		km	ft	min.	dial				
67	1	-13.6	400	89	3712.388				
68	2	-11.7	260	107	3719.501				
69	3	-9.7	340	139	3713.413				
70	4	-7.6	420	160	3706.677				
71	5	-4.7	260	156	3713.029				
72	6	-2.95	240	196	3711.941				
73	7	-0.7	360	212	3704.315				
74	base 1	0	400	41	3702.044				
75									
76									
77	***********								
78	elev.	drift	calibrate	convert	latitude	free air	Bouguer	from base	total
79	m	dial	mgal	gu	gu	gu	gu	gu	gu
80	=C67*FT_M	=E67-DRIFT*(D67-TB1)	=(B80-DIAL)*FACTOR+DIAL_MGAL	=C80*MGAL_GU	=D80-(LAT_COR*B67)	=E80+(FAC*A80)	=F80-(BC*A80)	=G80-G87	=H80+ABS
81	=C68*FT_M	=E68-DRIFT*(D68-TB1)	=(B81-DIAL)*FACTOR+DIAL_MGAL	=C81*MGAL_GU	=D81-(LAT_COR*B68)	=E81+(FAC*A81)	=F81-(BC*A81)	=G81-G87	=H81+ABS
82	=C69*FT_M	=E69-DRIFT*(D69-TB1)	=(B82-DIAL)*FACTOR+DIAL_MGAL	=C82*MGAL_GU	=D82-(LAT_COR*B69)	=E82+(FAC*A82)	=F82-(BC*A82)	=G82-G87	=H82+ABS
83	=C70*FT_M	=E70-DRIFT*(D70-TB1)	=(B83-DIAL)*FACTOR+DIAL_MGAL	=C83*MGAL_GU	=D83-(LAT_COR*B70)	=E83+(FAC*A83)	=F83-(BC*A83)	=G83-G87	=H83+ABS
84	=C71*FT_M	=E71-DRIFT*(D71-TB1)	=(B84-DIAL)*FACTOR+DIAL_MGAL	=C84*MGAL_GU	=D84-(LAT_COR*B71)	=E84+(FAC*A84)	=F84-(BC*A84)	=G84-G87	=H84+ABS
85	=C72*FT_M	=E72-DRIFT*(D72-TB1)	=(B85-DIAL)*FACTOR+DIAL_MGAL	=C85*MGAL_GU	=D85-(LAT_COR*B72)	=E85+(FAC*A85)	=F85-(BC*A85)	=G85-G87	=H85+ABS
86	=C73*FT_M	=E73-DRIFT*(D73-TB1)	=(B86-DIAL)*FACTOR+DIAL_MGAL	=C86*MGAL_GU	=D86-(LAT_COR*B73)	=E86+(FAC*A86)	=F86-(BC*A86)	=G86-G87	=H86+ABS
87	=C74*FT_M	=E74-DRIFT*(D74-TB1)	=(B87-DIAL)*FACTOR+DIAL_MGAL	=C87*MGAL_GU	=D87-(LAT_COR*B74)	=E87+(FAC*A87)	=F87-(BC*A87)	=G87-G87	=H87+ABS

Figure 3. Formulae comprising bottom part of template of Figure 2. Some of headers are truncated.

string by the leading "=". The entry LAT in cell F58 is a named constant which has been designated elsewhere to refer to cell B54, which contains the latitude of the base station in degrees. Spreadsheet programs include a number of intrinsic functions, such as the SIN function used in cell F58. Because trigonometric functions in spreadsheets generally operate on angles expressed in radians, the conversion of latitude from degrees to radians is made using the intrinsic function PI(), which is equal to π radians. The resulting latitude correction calculated for this base station latitude, 7.998 gu/km, is shown in cell F58 of Figure 2.

The block of cells under STATION INPUTS shows the utility of the spreadsheet format for data tabulation. The block under STATION CALCULATIONS illustrates the utility of the spreadsheet for repeated calculation. The equations entered in row 80 were copied down simply to rows 81-87. Relative cell references are changed automatically as the equations are copied. For example, the elevation conversion in cell A81 changes the elevation in feet for station 1 given in cell C67 to meters. As this equation is copied downwards for the other stations, the form of the equation and the constant remain the same, whereas the relative cell references are updated to refer to successive cells below C67. Absolute cell references can be used to avoid such updating, either by using labels as discussed, or using dollar marks with the cell designator (as in the equations in cells H80-H87).

Computer programs, similar to written language, should not just work but also be well-structured so that they are understandable, easily debugged, and easily modified (e.g., Kernighan and Plougher, 1978). The same can be said for spreadsheet style (Nevison, 1987). Figure 2 illustrates some aspects of good spreadsheet style: a general header, instructions for use, column headers, minimization of clutter, and the separation of cell blocks used for constants, data input, and output calculations.

Figure 4 is a spreadsheet-generated graph of the Bouguer anomaly given in cells I80-I86 of Figure 2. The data were collected as part of an exercise by a geophysics class along an approximately north-south profile in Lancaster County, Pennsylvania. The intent of the survey was to demonstrate to the class the latitudinal gradient of gravity. As the survey progressed and the data were being reduced and graphed, it became clear that as we moved north, Bouguer gravity was decreasing, contrary to what the latitude effect alone would yield. Thus, we were able to discuss during the survey the possible geological causes of this anomaly.

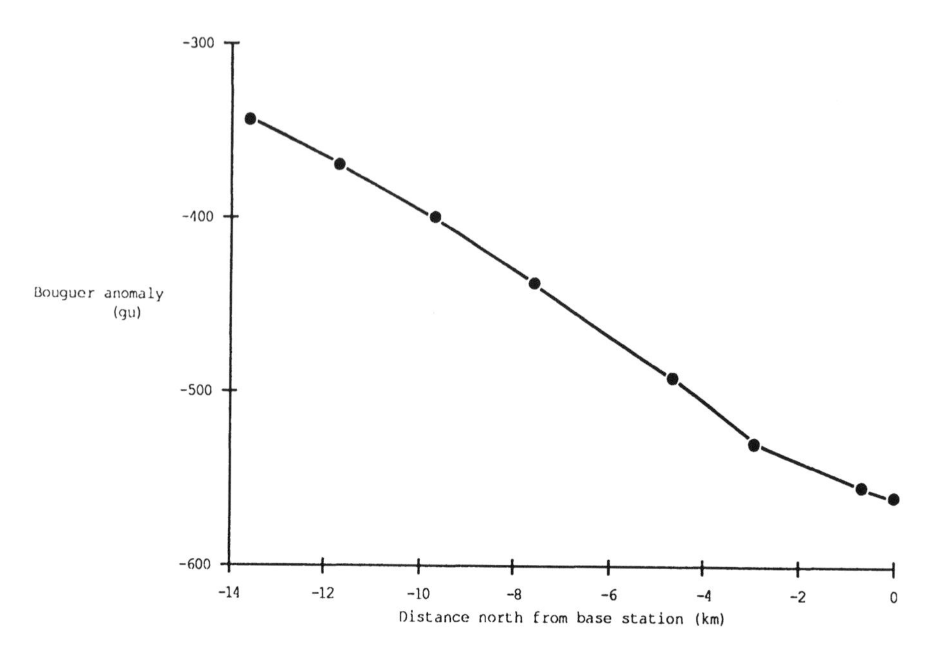

Figure 4. Bouguer anomaly results from spreadsheet of Figure 2.

Gravity - An Example of Data Modeling

The "what-if" capabilities of the spreadsheet make it an excellent tool for simple forward models of geophysical anomalies. In the spreadsheet, distance along the profile (independent variable) can be placed in one column, and the anomaly (dependent variable) can be calculated in another column. Properties of the body (e.g., size, depth, density contrast) can be set as constants in separate cells and easily changed. These changes immediately ripple through the rest of the spreadsheet, and the anomalies are recalculated for the new parameter values. The spreadsheet example in Figure 5 is for the model of a buried sphere, and includes some real data that can be modeled adequately with a sphere. Anomalies (cells C52-C72) are calculated at the positions on the profile (A52-A72) where data exist (column E), and at some additional positions in order to generate a smooth profile. Missing data in cells E52-E72 are indicated by a value of 999. Figure 6 shows the formulae used in this template. Figure 7 is a graph generated by the spreadsheet of the data and a model anomaly, using the parameters shown in Figure 5. As the model parameters in the spreadsheet template are changed, the graph also is updated automatically, facilitating iteration towards a good-fitting model.

	A	B	C	D	E	F	G
1							
2			GRAVITY MODELING FOR A BURIED SPHERE				
3							
4	**						
5							
6	Name:	Grav-Sph					
7							
8	Author:	Rob Sternberg					
9		Franklin & Marshall College					
10	Date:	Sept-86					
11							
12	Purpose:	To calculate the gravity anomaly along a traverse above a					
13		buried sphere, and to plot model anomaly.					
14		Data can be plotted on same graph, but only one series of					
15		distances can be entered. Missing values can be used					
16		in either modeled anomaly or data, but will plot as zeros.					
17							
18	Instructions:						
19		1 Enter any notes on survey data.					
20		2 Enter radius, depth, and density contrast of body.					
21		3 Enter distances for which anomaly is to be calculated.					
22		4 Copy anomaly formula to corresponding cells.					
23		5 Enter anomaly data.					
24		6 Adjust graph parameters as necessary.					
25							
26	References:		Introduction to Geophysical Prospecting (3rd ed.)				
27			Milton B. Dobrin				
28			McGraw-Hill, New York, 1976, ch. 11				
29							
30	Modified:	Jun-89	from Macintosh/Excel to NEC/Quattro				
31							
32	**						
33							
34	DATA NOTES						
35	Data from L. L. Nettleton						
36	Taken from Turcotte and Schubert, Fig. 5-10						
37	Salt dome 125 miles SE of Galveston, TX						
38							
39	INPUT PARAMETERS FOR SPHERE				UNIVERSAL CONSTANT		
40	radius	4000	meters		G	6.670E-11	
41	depth	6000	meters		4/3*PI*G	2.79E-10	m3/(kg s2
42	dens cont	-200	kg/m3				
43					RUN CONSTANTS		
44					max anom	-99.34	gravity u
45					1/2 width	4598	meters
46							
47							
48	MODEL				DATA		
49	******************************				*********		
50	Distance	Distance	G anomaly		G anomaly		
51	meters	km	gu		gu		
52	-10000	-10	-14		999.00		
53	-8500	-9	-19		-11.10		
54	-7750	-8	-23		-19.40		
55	-6875	-7	-28		-30.60		
56	-6000	-6	-35		-40.30		
57	-5250	-5	-42		-50.00		
58	-4375	-4	-52		-61.10		
59	-3500	-4	-64		-70.80		
60	-1500	-2	-91		-80.60		
61	-1000	-1	-95		999.00		
62	0	0	-99		999.00		
63	1000	1	-95		-90.30		
64	1500	2	-91		999.00		
65	3375	3	-66		-80.60		
66	4000	4	-57		-70.80		
67	4750	5	-48		-61.10		
68	5375	5	-41		-50.00		
69	6125	6	-34		-41.70		
70	7250	7	-26		-30.60		
71	8125	8	-21		-20.80		
72	10000	10	-14		-11.10		

Figure 5. Spreadsheet template for modeling of gravity anomaly over buried sphere. Included are some data that can be represented adequately by this model. Data are from Figure 5-10 of Turcotte and Schubert (1982).

	A	B	C	D	E	F	G
34	DATA NOTES						
35	Data from L						
36	Taken from						
37	Salt dome 1						
38							
39	INPUT PARAM				UNIVERSAL CO		
40	radius	4000	meters		G	0.0000000000667	
41	depth	6000	meters		4/3*PI*G	-4/3*PI()*G	m3/(kg s2)
42	dens contra	-200	kg/m3				
43					RUN CONSTANT		
44					max anom	-(F41*RHO*R ^3/Z^2)*1000000	gravity units
45					1/2 width	-0.7663*Z	meters
46							
47							
48	MODE				DATA		
49	***********				*********		
50	Distance	Distance	G anomaly		G anomaly		
51	meters	km	gu		gu		
52	-10000	-A52/1000	-GMAX/(1+(A52/Z)^2)^1.5		999		
53	-8500	-A53/1000	-GMAX/(1+(A53/Z)^2)^1.5		-11.1		
54	-7750	-A54/1000	-GMAX/(1+(A54/Z)^2)^1.5		-19.4		
55	-6875	-A55/1000	-GMAX/(1+(A55/Z)^2)^1.5		-30.6		
56	-6000	-A56/1000	-GMAX/(1+(A56/Z)^2)^1.5		-40.3		
57	-5250	-A57/1000	-GMAX/(1+(A57/Z)^2)^1.5		-50		
58	-4375	-A58/1000	-GMAX/(1+(A58/Z)^2)^1.5		-61.1		
59	-3500	-A59/1000	-GMAX/(1+(A59/Z)^2)^1.5		-70.8		
60	-1500	-A60/1000	-GMAX/(1+(A60/Z)^2)^1.5		-80.6		
61	-1000	-A61/1000	-GMAX/(1+(A61/Z)^2)^1.5		999		
62	0	-A62/1000	-GMAX/(1+(A62/Z)^2)^1.5		999		
63	1000	-A63/1000	-GMAX/(1+(A63/Z)^2)^1.5		-90.3		
64	1500	-A64/1000	-GMAX/(1+(A64/Z)^2)^1.5		999		
65	3375	-A65/1000	-GMAX/(1+(A65/Z)^2)^1.5		-80.6		
66	4000	-A66/1000	-GMAX/(1+(A66/Z)^2)^1.5		-70.8		
67	4750	-A67/1000	-GMAX/(1+(A67/Z)^2)^1.5		-61.1		
68	5375	-A68/1000	-GMAX/(1+(A68/Z)^2)^1.5		-50		
69	6125	-A69/1000	-GMAX/(1+(A69/Z)^2)^1.5		-41.7		
70	7250	-A70/1000	-GMAX/(1+(A70/Z)^2)^1.5		-30.6		
71	8125	-A71/1000	-GMAX/(1+(A71/Z)^2)^1.5		-20.8		
72	10000	-A72/1000	-GMAX/(1+(A72/Z)^2)^1.5		-11.1		

Figure 6. Formulae comprising bottom part of template of Figure 5. Some of headers are truncated.

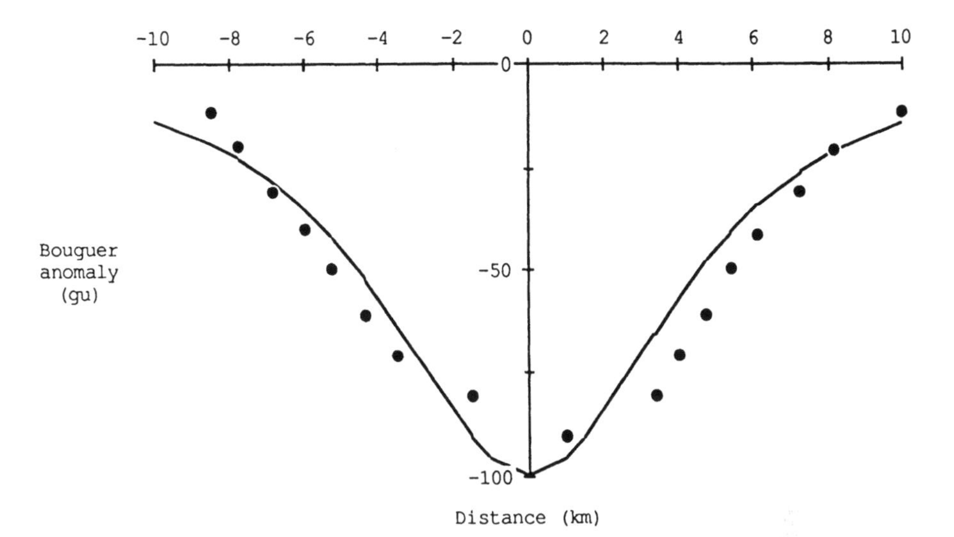

Figure 7. Graphs of data (filled symbols) and model anomaly (solid line) from template of Figure 5.

Other Applications

The examples given here indicate a few of the many possible applications of laptop computers and spreadsheets in geophysical surveying. Some other possibilities are suggested here.

(1) Calculating gravity anomalies for other bodies with simple geometry. Other simple geometries, such as the buried horizontal cylinder, are also easy to set up on a spreadsheet. Modeling of polygonal bodies would be possible, although more complex.

(2) Uploading of proton precession magnetometer data. Many models now have digital memories and serial interfaces which can be connected to the serial port on the host computer. Uploading can be accomplished with a simple BASIC program, or with vendor-supplied software. Data collection in small surveys can be rapid, and the magnetometer memory may be filled in several hours, requiring it to be dumped.

(3) Drift correction for magnetometer data. This is especially easy for simple survey designs, such as base station reoccupations before and after each line is run. This is feasible for small surveys. The corrected profiles then can be plotted while in the field.

(4) Calculating magnetic anomalies. Again, this is straightforward for simple geometries.

(5) Calculating apparent resistivities. The spreadsheet is useful for logging the data acquired in resistivity profiling or sounding. The readings then can be converted into apparent resistivities by using the appropriate equation and array parameters. Resistivity profiles and sounding curves can also be plotted.

(6) Logging/acquisition of electromagnetic survey data. Electromagnetic conductivity surveys can proceed quite. For a small-scale survey, one member of the crew can make the readings, and another can manually log the data into the computer. Some meters have been modified so that readings are triggered with a switch, and transmitted to a digital data logger. These readings can then be uploaded periodically to the host computer. Profile plots can be generated in the field.

ACKNOWLEDGMENTS

Several people and institutions helped to make this paper possible. My geophysics students stimulated me to develop these ideas, and helped to test them. Steve Spadafore of Franklin and Marshall College designed and constructed the seismograph interface, and wrote the heart of the BASIC program that made seismogram trace acquisition possible. Justine Fluck helped to prepare my poster for the 1989 International Geological Congress which formed the basis for this article. Franklin and Marshall College provided various types of support. The University of Konstanz (Konstanz, Federal Republic of Germany) provided a place to write this article during my sabbatical. The paleomagnetics laboratory at Eidgenossiche Technische Hochschule, Zürich, kindly allowed me to use their laser printer for manuscript preparation.

REFERENCES

Dory, R.A., 1988, Spreadsheets for physics: Computers in Physics, v. 2, no.3, p. 70-74.

Guglielmino, R., 1989, Using spreadsheets in an introductory physics lab: The Physics Teacher, v. 27, p. 175-178.

Herweijer, J.C., 1986, Lotus 123 - a graphical spreadsheet program used for geophysical calculations: First Break, v. 4, no. 6, p. 9-13.

Jones, D., 1983, Geophysical data handling in the field using microcomputer techniques: Paper presented at the meeting of the Society for Exploration Geophysicists, Las Vegas, Nevada.

Kelly, M.A., Dale, P., and Haigh, J.G.B., 1984, A microcomputer system for data logging in geophysical surveying: Archaeometry, v. 23, p. 169-188.

Kernighan, B.W., and Plougher, P.J., 1978, The elements of programming style (2nd ed.): McGraw-Hill Book Co., New York, 168 p.

Nevison, J.M., 1987, The elements of spreadsheet style: Prentice-Hall, New York, 197 p.

Spadafore, S., and Sternberg, R.S., 1990, A serial interface for digital communications between a Bison one-channel signal-enhancement seismograph and a host computer, in preparation.

Sutter, T.C., and Sternberg, B.K., 1988, Gravity survey data reduction using Symphony, Concerto, and QuickBasic: COGS Computer Contributions, v. 4, p. 1-36.

Tarkoy, P.J., 1986, Application of microcomputers and generic software in engineering geology: Bull. Association Engineering Geologists, v. 23, p. 221-242.

Turcotte, D.L., and Schubert, G., 1982, Geodynamics - applications of continuum dynamics to geological problems: John Wiley & Sons, New York, 450 p.

Walter, R.K., 1989, Simulating physics problems with computer spreadsheets: The Physics Teacher, v. 27, p. 173-175.

A PROGRAM FOR PETROPHYSICAL DATABASE MANAGEMENT

Bernhard B. Holub
Leoben University, Leoben, Austria

ABSTRACT

The personal computer program ROCKBASE based on dBASE IV is presented. The aim of this menu-driven and mask-oriented interactive program is the management of data from irregularly distributed field samples that are measured in the laboratory. Statistical modules such as calculation of mean values are offered, but the main purpose of ROCKBASE is the preparation of the data for graphic output. As an interface the ASCII format enables full compatibility with commercial graphic packages. It is shown that thematic maps of an investigation area can be produced with little expense.

INTRODUCTION

The measurement of petrophysical parameters is of great importance for many problems in geoscience. These parameters mainly comprise seismic velocities, specific electrical resistivity and polarisability, magnetic susceptibility, intensity and direction of natural remanent magnetization (NRM), rock density, and natural radioactivity. The investigations concern both 'soft' rocks, 'hard' rocks, and artificial deposits (dams, waste deposits, dumps) for prospecting, geotechnical and hydrogeological, as well as environmental problems.

Use of Microcomputers in Geology, Edited by D.F. Merriam
and H. Kürzl, Plenum Press, New York, 1992

Depending on the problem and the measuring technique, the data will be collected in the field, using gridded or irregularly distributed points, in the laboratory from hand samples, or measured directly by well logging. For easy administration of such amounts of data, a database-management system on a mainframe or personal computer is normally used. In contrast to many highly developed programs for well-logging data, individual solutions, designed especially for irregular distributed measuring data from the field, generally are utilized (Puranen and Hongisto, 1989).

Because of the growing importance of petrophysics, several papers have been published within the last years, that summarize the petrophysical parameters of important minerals and rocks (Kobranova, 1989; Puranen, 1989; Puranen, Elo, and Airo, 1978; Schön, 1983; Touloukian, Judd, and Roy, 1981).

In this paper, the personal computer program ROCKBASE is presented, which depends on the database system dBASE IV. It is designed for management of petrophysical data generated in the laboratory from field samples. The database program is adaptable easily for other applications such as hydrogeology, geochemistry, or petrography.

FUNDAMENTALS

dBASE IV is a relational database-management system with its own language, that enables the creation of specific programs, and speeds up routine work. The programs run under the control of dBASE IV, or independently using a runtime module. The database structure of dBASE IV contains a flexible number of equivalent data fields with defined length, which can be treated and presented in any order and selection. The individual data fields can be defined as alphanumeric, numeric, logical, or data fields. Each database can be sorted in ascending or descending order on the basis of an index field. Using a selection criteria, a restricted part of the database can be viewed, or written to an ASCII file.

PROGRAM ORGANIZATION

The structure of ROCKBASE is divided into a statistical part, and a part containing the measurements. The statistical part contains information about

- Sample name
- Rock type
- Rock code
- Sample locality
- Map sheet
- Coordinates of sample locality
- Date of sample collection
- Sample orientation
- Name of sample collector
- 2 fields for individual selection criteria
- Measuring method
- Comment

In the other part, the measurements are stored in SI-units in the same file as the statistical data, and comprise the following data fields:

- Density [kg/m^3]
- Susceptibility [10^{-3}]
- Resistivity [Ω.m]
- IP-effect [%]
- Intensity of NRM [nT]
- Declination of NRM [°]
- Inclination of NRM [°]

Additional data fields, for example seismic velocities, can be appended easily to the database. On the other hand, unused fields can be removed.

ROCKBASE allows a completely menu-driven data management (Fig.1) with little knowledge about database structure and syntax rules. For complex questions a direct data access with dBASE query language is possible additionally to the program.

After selection of the suitable database in the menu 'Database/Select' one can determine a data field in 'Mode/Sort' after which the database is sorted. In addition to the default sorting by sample name, one can select rock code, map sheet, or measurement values.

An input mask in 'Mode/Filter' enables the construction of selection criteria by fixed numerical values (e.g. map sheet 163) and also with help of comparison operators (e.g. density >3000). Complex selection criteria also can be defined using dBASE query language. The total number of records in the database, and the number of those records which match the selection criteria, are displayed on the screen (Fig.2). The defined

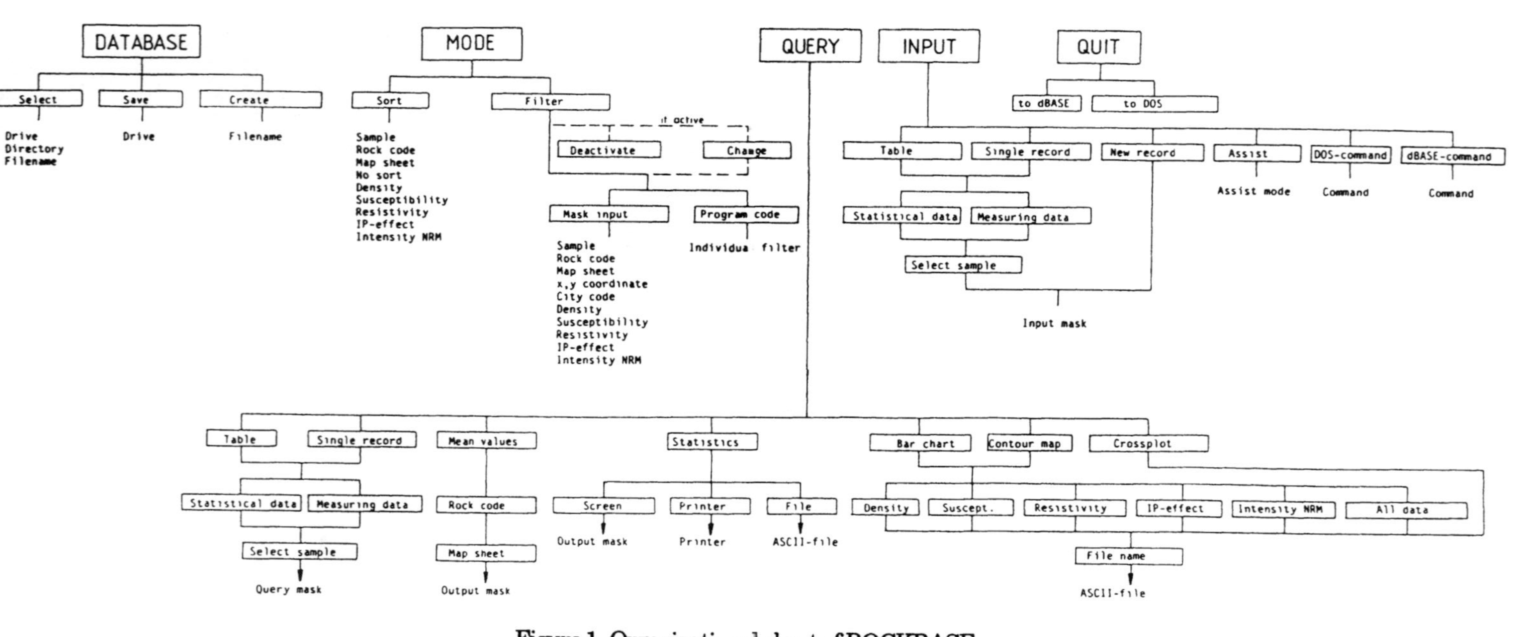

Figure 1. Organizational chart of ROCKBASE.

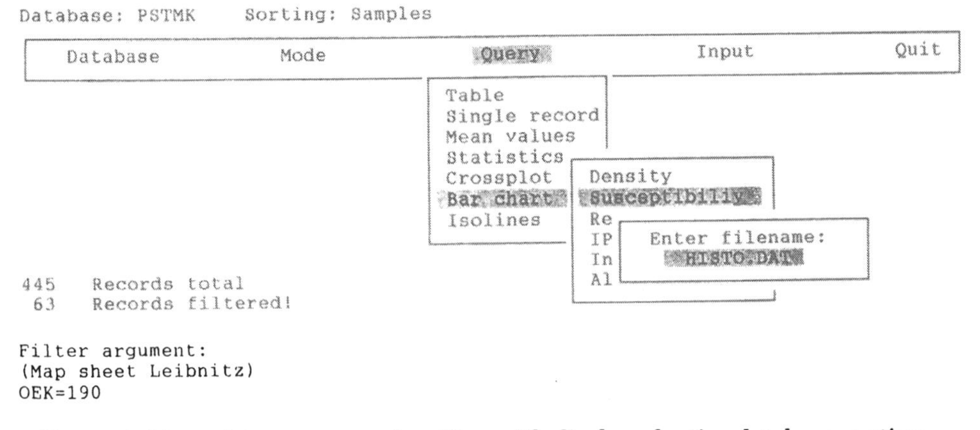

Figure 2. Menu-driven program handling with display of active database, sorting criterion, total number of records, and selection criteria.

criteria remain active for all subsequent program actions, but are changeable or could be deactivated at any time.

Data input or alteration is done by a mask for statistics and measurement values (Fig.3). Either the mode 'Input/Single record' - for just one data record displayed - or 'Input/Table' - for several records listed in a table can be selected.) These masks are organized in that way, such that misentries are avoided - for example, an 'empty' entry in the field of sample name or, in numerical fields, unrealistic values such as density <1000. If no measurement data are available, the appropriate data fields are marked by a negative number (e.g. resistivity -1). The same masks used for data input are available for data query with the difference, that the data are only readable to prevent misentry.

In 'Query/Mean values' the means for each measured parameter of a special rock type which is represented by its rock code, are calculated. Those map sheets on which the selected rock type occurs are displayed on the screen and one can select one or more map sheets to calculate the mean values. The output mask shows the number of measuring values, minimum, maximum, and the mean value with standard deviation for each parameter (Fig.4).

In the menu 'Query/Statistics', all rock types that match the selection criteria are displayed in conjunction with the total number of stored samples and the number of measured values for density, susceptibility, etc. (Fig.5). The output can also be directed to a printer or written to an ASCII file.

A main part of ROCKBASE is the data output to ASCII files. So the measured values can be transferred easily to graphic programs, or

```
Sample     KD06.4                                              A
Rock       Marble gray, banded
Locality   Q Haider, 1km NE Freiland
Rock code  ST085

Map sheet 189

           X-Koord     659760          Date      [illegible]
           Y-Koord     191230          Name      [illegible]
           City code    60306          Orient.   [illegible]

           Comment  Literature: Beck (1984,1987)

Code1 X

Code2
```

```
            Sample  L03              ST064                                   B
                                     Phyllite graphitic, ore bearing
           Density   2872
    Susceptibility     1.450         Main road 300m NW St.Georgen,km 7.0
       Resistivity    246            Map sheet  159
         IP-effect      3.32
     Intensity NRM     -1
  Measuring method      A
   Additional data      F

Code1 B

Code2
```

Figure 3. Input mask for single record mode. Misentry is prevented by input checking. A, Statistical data; B, Measuring data.

```
ST081       Gneiss flattened           Crystalline Basement
            Map sheet: 188 189 205 206  (all)
            51 Samples

Density:                                    Resistivity:
n=51              [kg/m3]                   no data
Minimum            2548
Maximum            2973
Mean value         2818 ± 81

Susceptibility:                             IP-effect:
n=47             [10-3 SI]                  no data
Minimum           0.220
Maximum           1.170
Mean value        0.570 ± 0.204
```

Figure 4. Calculation of minimum, maximum, mean values, and standard deviation of selected rock type with display of map sheets of its occurrence.

Database: PSTMK.DBF
Filter argument:
OEK=189
(Map sheet Deutschlandsberg)

Code	Rock type	n	Den	Sus	Res	IP	Int
ST081	Gneiss flattened	38	38	37	-	-	-
ST082	Pegmatoid gneiss and micaschist	16	16	13	-	-	-
ST083	Pegmatoid gneiss	13	13	13	-	-	-
ST084	Eclogite, metagabbro	6	6	6	-	-	-
ST085	Marble	6	6	5	-	-	-
ST088	Quartzite	4	4	3	-	-	-
ST092	Amphibolite	2	2	2	-	-	-
ST094	Paragneiss, Plagioklasegneiss	13	13	11	-	-	-
ST095	Pegmatite	3	3	2	-	-	-

Figure 5. List of all rock types with total amount of samples and number of measuring values for each petrophysical parameter.

treated as tables. In the menu 'Query' one can select between output for crossplots, bar charts, or contour maps (Fig.2).

Additional to the listed possibilities, dBASE or DOS commands can be used within ROCKBASE. On the other hand, the 'assist mode', the user interface of dBASE IV can be accessed.

In the menu 'Database/Save', the active database will be written to disk. New databases are created by the database structure which is stored in a control file.

GRAPHIC OUTPUT

For further graphic processing, the filtered data are stored in ASCII format. This format is readable by any program and offers many possibilities for individual use of graphic software (Fig.6). Many graphic programs enable the creation of output masks (e.g. semilogarithmic crossplots with axes labels and grid lines) and storage in control files. With a batch file, the ASCII data from ROCKBASE are included in those masks automatically. In this way a graphic output with different data groups is easily achieved.

Crossplots are the most important type of presentation for petrophysical data. These diagrams allow comparison of all possible parameters (e.g. density vs. susceptibility). Using X and Y coordinates, position maps of sample localities also can be established.

Bar charts are used to show the frequency distribution of parameters. The necessary calculation is done easily in spreadsheet programs such as Lotus 1-2-3.

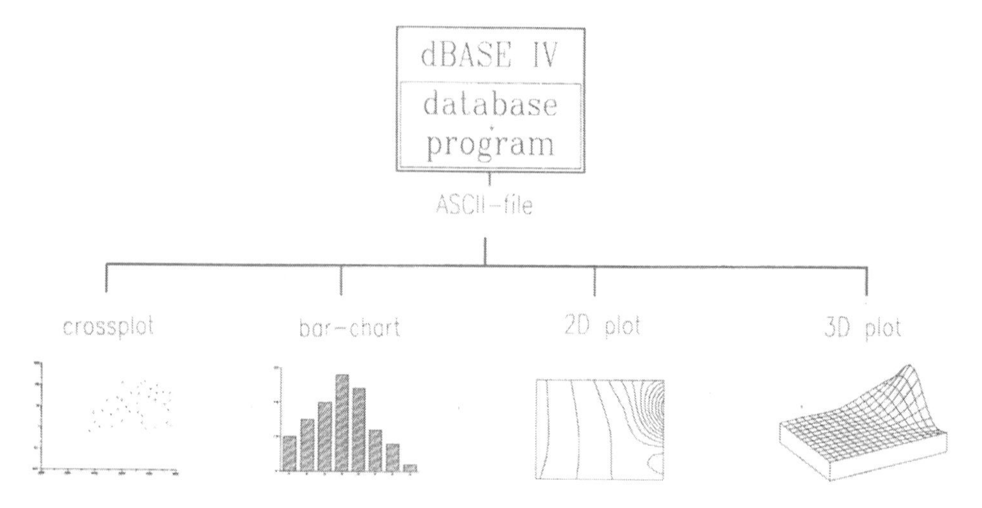

Figure 6. Possibilities of graphic processing of output files created by ROCKBASE.

Contour maps are used to provide surface distribution of rock parameters in two or three dimensions. This consideration allows the delimitation of anomalies, or the investigation of surface inhomogeneities of a rock type. In contour mapping it is important to consider the method of grid calculation from irregular distributed data. As well developed and powerful tool to create 2D and 3D plots, the SURFER package was proved. This program enables graphic output in DXF format, which offers further processing in CAD programs such as AutoCAD. In that way digital thematic maps of a measuring area with additional topographic information (cities, river net, digital surface model, etc.) can be produced with little effort from the data managed by ROCKBASE.

ACKNOWLEDGMENTS

I wish to thank R. Puranen, H. Hongisto, and L. Kivekäs from the Geophysics Department of the Geological Survey of Finland in Espoo/ Otaniemi for constructive discussions during my academy exchange in May 1989.

This work was supported financially by the Austrian Academy of Sciences' programme 'Geophysik der Erdkruste'.

REFERENCES

Kobranova, V.N., 1989, Petrophysics, Springer-Verlag, Berlin, Heidelberg, New York, 375 p.

Puranen, R., 1989, Susceptibilities, iron and magnetite content of Precambrian rocks in Finland: Geol. Survey Finland, Rept. Invest. 90, 45 p.

Puranen, R., Elo, S., and Airo, M.-L., 1978, Geological and areal variation of rock densities, and their relation to some gravity anomalies in Finland: Geoskrifter, v. 10, p. 123-164.

Puranen, R., and Hongisto, H., 1989, Petrofysikaalisten tietorekistereiden kehitys. Summary: Development of petrophysical databases: Geol. Survey Finland, Interim Report Q12/27/89/1, 10 p.

Schön, J., 1983, Petrophysik, Physikalische Eigenschaften von Gesteinen und Mineralen: Akademie-Verlag, Berlin, 405 p.

Touloukian, Y.S., Judd, W.R., and Roy, R.F., 1981, Physical properties of rocks and minerals, *in* Touloukian, Y.S., and Ho, C.Y., eds., McGraw-Hill/CINDAS Data Series on Material Properties, v. II-2: McGraw-Hill Book Co., New York, 548 p.

ARTIFICIAL INTELLIGENCE TOOLS AND TECHNIQUES FOR WATER-RESOURCES ASSESSMENT IN AFRICA

Patrice Poyet
Centre Scientifique et Technique Du Batiment, Valbonne, France

and

Michel Detay
International Training Centre for Water Resources Management, Valbonne, France

ABSTRACT

This paper presents the development of a hydrogeological expert system able to handle the drilling location problem within the scope of village water-supply programs. This work is based on the experience the authors gained from thousands of drillings carried out in fifteen African countries. The cognitive model comes from the practical know-how acquired from real-world programs, from original statistics and probabilistics analyses showing connections between data collected during the drilling-site selection and hydrodynamic parameters registered in the borings and from the research the authors carried out in the artificial intelligence field to propose a comprehensive knowledge modeling framework. The paper includes a description of the specific knowledge involved in the drilling location process. Relevant hydrogeological parameters recognition and examples of advanced computer knowledge modeling methods are presented. First the rules of thumb and the interpretative frames retained for the cognitive model are described, then the characteristics of the HYDROLAB expert system devoted both to computer-aided decision and to computer-assisted learning support.

Use of Microcomputers in Geology, Edited by D.F. Merriam and H. Kürzl, Plenum Press, New York, 1992

INTRODUCTION

The various estimations carried out within the scope of the International Drinking Water Supply and Sanitation Decade (1981-1990) show that the water demand in countries of Western and Central Africa is considerable. The most optimistic statistics indicate that drinking water is not available to more than 80% of the rural population in developing countries. The problem of water prospecting is more than ever a crucial objective. Even if African states have devoted themselves for more than twenty years to the improvement of the water supply to the rural population, extrapolation of the 1976 estimations suggest that 100,000 water points need to be created by 1990. To meet the water demand in developing countries, thousands of water points remain to be created within the scope of village water-supply programs.

Since 1973, we have been working on the question of extending the water resources of developing countries and we have analyzed the mechanisms governing the hydrodynamic characteristics of aquifers in drought-prone areas. The considerable wealth of knowledge and experience we gained from the aforementioned work, the existence of a large amount of statistical data, the repetitive nature of the steps to be carried out for a hydrogeological search and the African technicians' training needs, enabled us to develop a hydrogeological expert-system consultant.

This paper is an overview of hydrogeological studies in the field of village water-supply programs. Its primary interest lies in the way it identifies decisive hydrogeological parameters and in its use of artificial-intelligence techniques in this domain. We first present the main features of African hydrogeology which represent the context of this study, then we summarize the methods used by experts for well-location studies. These techniques and the specialists' behavior represent the background we analyzed to recognize a suitable cognitive material for the knowledge modelling process we aimed to achieve. We explain our methodology to identify decisive hydrogeological parameters and present original data showing connections between data collected during the location studies and the hydrodynamic parameters registered when operating the drillings. The HYDROLAB expert-system consultant is the concrete result of this research, as it embodies the previously recognized rules of expertise, their associated interpretative models, and the related decision-making parameters. We finally present some insight to the expert-system architecture and describe its range of application, either as a computer-based expert consultant to help in solving the well-site location, or as a computer assistant for village water-supply courses.

GEOLOGICAL CONTEXT OF THE STUDY

Groundwater usually is connected directly to geological units, and this remark takes its full significance in village water supply as the strategies involved widely rely on the recognized geological context. In Africa, and following a schematic description, the geological context is divided into three large groups: the Precambrian crystalline bedrock, corresponding to the basement complex that represents the oldest formations of the African Shield (granites, gneisses, quartzites, schists, etc.); old formations, tabular, Infracambrian and Primary, associated with the deep-seated complex; and sedimentary formations, both post-primary and recent deposits. From these three geological contexts, two major types of aquifer systems can be determined: sedimentary and igneous/metamorphic.

The sedimentary formations and recent deposits are permeable formations and generally enclose continuous aquifer. These are encountered in large sedimentary basins, mainly in the Sahel latitude: the Senegalo-Mauritanian basin, the central delta of the Niger river, Taoudeni Basin, Nigerian Basin, and Chadian Basin. They represent 85% of the surface area of Senegal, 65% of Mauritania, 75% of Nigeria, 64% of Mali, 65% of the Congo, 52% of Chad, and form a narrow costal band occupying a small portion of the Ivory Coast, Togo, Benin, Gabon, the Togo and Cameroon. The Precambrian crystalline bedrock and the old formations characteristically have a discontinuous aquifer. They occupy most of the Ivory Coast (97% of the country's surface area), Burkina Faso (95%), Togo (94%), Benin (83%), Cameroon (89%), and Gabon (80%), and correspond to zones of high population density.

The research that we have undertaken in this field since 1973, covers fifteen or so countries in Africa (Bernardi and Mouton, 1975; Bernardi and Detay, 1989; Detay, 1987; Detay and Poyet, 1988a, 1988b, 1989a, 1989b, 1990; Detay and others, 1986, 1990) and the resolution techniques proposed should address the continental scale of the problem. Few tasks in applied hydrogeology are more difficult than locating drilling sites for water in igneous and metamorphic rocks and 50% of the wells in some areas are registered as failures. Extreme variations of lithology and structure coupled with highly localized water-producing zones make geological and geophysical exploration difficult, especially for the Basement Context. Soil and vegetation usually cover outcrops and make the necessary detailed geological observations impossible.

We would remind the reader that rural water-supply programs aim at delivering good quality water but the expected yields are relatively low as a cubic meter per hour can be considered a satisfying result. The main difficulties concern the well-site selection for which no formal approach is available. These tasks require a great amount of expertise and involve conceptual and cognitive models as well as statistical studies that summarize in some way the knowhow gained from the practical programs accomplished.

The steps followed by experts during the interpretative process and the decisional variables involved to determine the location strategy are drastically different for sedimentary formations and for the crystalline environment. Sedimentary contexts involve a dedicated approach for each formation submitted according to its observed hydrogeological characteristics, whereas a generic behavior can be modeled in the situation of a saprolite reservoir enclosing a discontinuous aquifer. It is worthwhile to notice that our attention will be focused within the scope of this paper on the methodology to be followed for the crystalline basement context, as a general reasoning frame of the cognitive processes involved can be modeled in this situation.

DEVELOPMENT OF A COGNITIVE MODEL

Introduction

The first step in the modeling of a cognitive process is to identify the collection of data involved in the activity and a set of mechanisms that should be triggered when some conditions are fulfilled. Elementary information will be referred to as basic data, and will be used to build decisional variables of different orders thanks to an analysis of the tasks carried out by the specialists. A set of specialized inference rules aims to reproduce some intelligent behaviors triggered when their premises are satisfied and accounts for the solution recognition phase. We describe here the main phases of the cognitive process including the gathering and the selection of basic data, the recognition of decisional variables and of their associated interpretative models, the description of a typology of the various sets of inference rules required to account for knowledge-based activities, a short presentation of the solution models we have developed, and a summary of the decision making parameters used both to provide in-situ drilling strategy and to forecast some hydraulic parameters for the future wells. Figure 1 is a brief overview of the overall strategy we have designed to account for the knowledge involved in selecting well-site. One should notice that the process is truly expertise based and that sophisticated mathematical models are seldom involved as they usually require detailed calibration sets of data not available for rural studies in Africa and as they lead to an over-accurate local knowledge that is not reusable for further large-scale investigation even for rather similar sites showing an inherently different calibration set. To avoid the problems caused by such computational sensitivity, we followed a statistical approach enabling us to forecast some of the estimated characteristics of the drillings thanks to a set of robust decision-making-parameters recognized by experts and accurately modeled by statistical means.

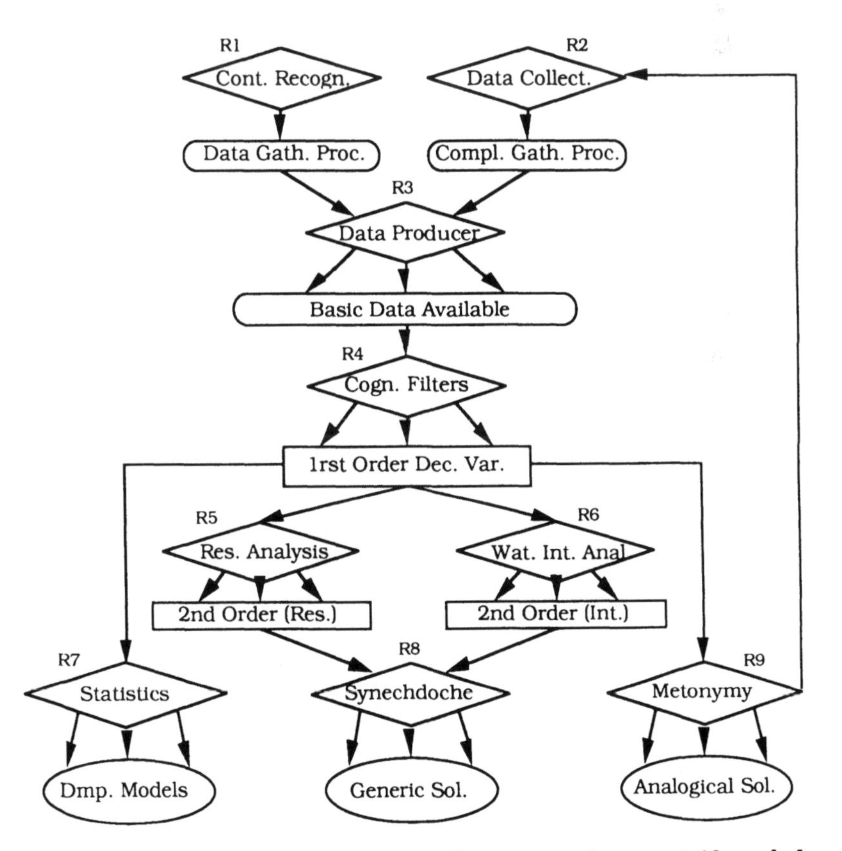

Figure 1. Overview of cognitive model proposed to account for experts' knowledge and activities involved in handling drillings location problem. Rules are indicated by diamond-shaped boxes, data or decisional variables by rectangles or smoothed rectangles (round-ended), solution models and decision-making parameters by ellipses.

The Assessment Phase

Even if the behavior of each expert is idiosyncratic, general guidelines can be proposed to conduct the data-collecting phase according to the synthesis of the observed attitudes. To amass the necessary elementary data, groundwater development normally is preceded by proper investigation and assessment and four major techniques contribute to a rational groundwater exploration. Each of these produce basic data that need to be combined within a semantic network to furnish meaningful first-order decisional variables allowing the generation of a valuable insight to the geological context.

Basic studies, such as the localization of villages on maps and inventory of essential data easily available such as rainfall measures, pluviometric deficits, catchment basin surface area, estimation of the runoff, geomorphological position, etc., are the first steps to be carried out and provide interesting primary data. There are completed by the study of aerial photographs which make excellent base maps. Stereopairs can be used for three-dimensional study of hydrologic features in order to distinguish rock and soil types, identify pattern of fractures, detect springs and marshes, all of them being considered as elementary information.

Then, field studies allow on-site recognition of relevant geological features. At this stage an inventory of water resources can be carried out to show the expert other fundamental parameters such as the thickness of the weathered zone or the position and variations of the watertable level. Moreover a discussion with the village authorities will allow the socioeconomic demands of the village to be taken into account. Finally geophysical exploration methods are used to obtain complementary material on the character of formations. The electrical resistivity method is a major geophysical tool used in groundwater exploration efforts and the results derived from such studies will reinforce the basic set of data collected.

When all these tasks have been scheduled (Data Gathering Process on Figure 1), the specialist is confronted with a set of intricate data (Basic Data illustrated on Figure 1), some of them being highly relevant for the current situation and some others presenting only a relatively low significance. It is worthwhile to select among the available information those meaningful for the current context (i.e., first order decisional variables) and to exhibit a cognitive model that will enable us to handle more general concepts such as the quality of the groundwater intake or the interest of the reservoir (i.e., secondary order variables).

It helps the later classification of a region into areas of good, fair, and poor groundwater prospects and represents a basis of the decision-making process. The HYDROLAB system developed corresponds exactly to such an approach and we shall see that it reproduces the data collecting phase as a task scheduling process in the computer science sense, and that the system's attention is focussed on the relevant decisional variables and associated interpretative models to achieve the matching with analogical solutions and the generation of second-order variables suitable for the recognition of the generic solutions. Finally the results are expressed in terms of decision-making parameters giving help to the program manager.

Among the aforementioned elementary data issued from the gathering phase, we propose now a comprehensive evaluation of the first-order decisional variables used by groundwater specialists to perform well-location.

Search for Decisive Hydrogeological First-Order Variables

Among the many elementary data previously collected, not all of them are equally interesting for the Basement Context we wish to model here. Favorable ones are identified as connection tropes, as each of them can be considered as valuable for the assessment of the overall context, and obey a mental approach close to synecdoche, a process permitting to establish a connection from some part to the whole.

As such, we extracted from this set ten first-order decisional variables believed to be the major meaningful hydrogeological parameters for the crystalline aquifer environment. Each of these first-order variables can be used by experts as a partial but significant appraisal of more general concepts such as the amount of water intake or the potential of the water-bearing rocks, and represent different parts useful to address a wider objective later on described as second-order decisional variables. For each of these connection tropes, we implemented for the automated system expertise rules (i.e. the inferential or decisional process of the experts) and interpretative models (i.e. mental representation or statistical models of the analyzed process).

The first-order decisional variables that have been retained for the model presented are inferred from the basic data (refer to Fig. 1). Figure 2 illustrates some of the associated rules of expertise (referred to as Re.) and of the connected interpretative models (referred to as Im.) used for such a processing. It also shows the general logic involved by the program to help the decision-making process, highlighting the data required, the

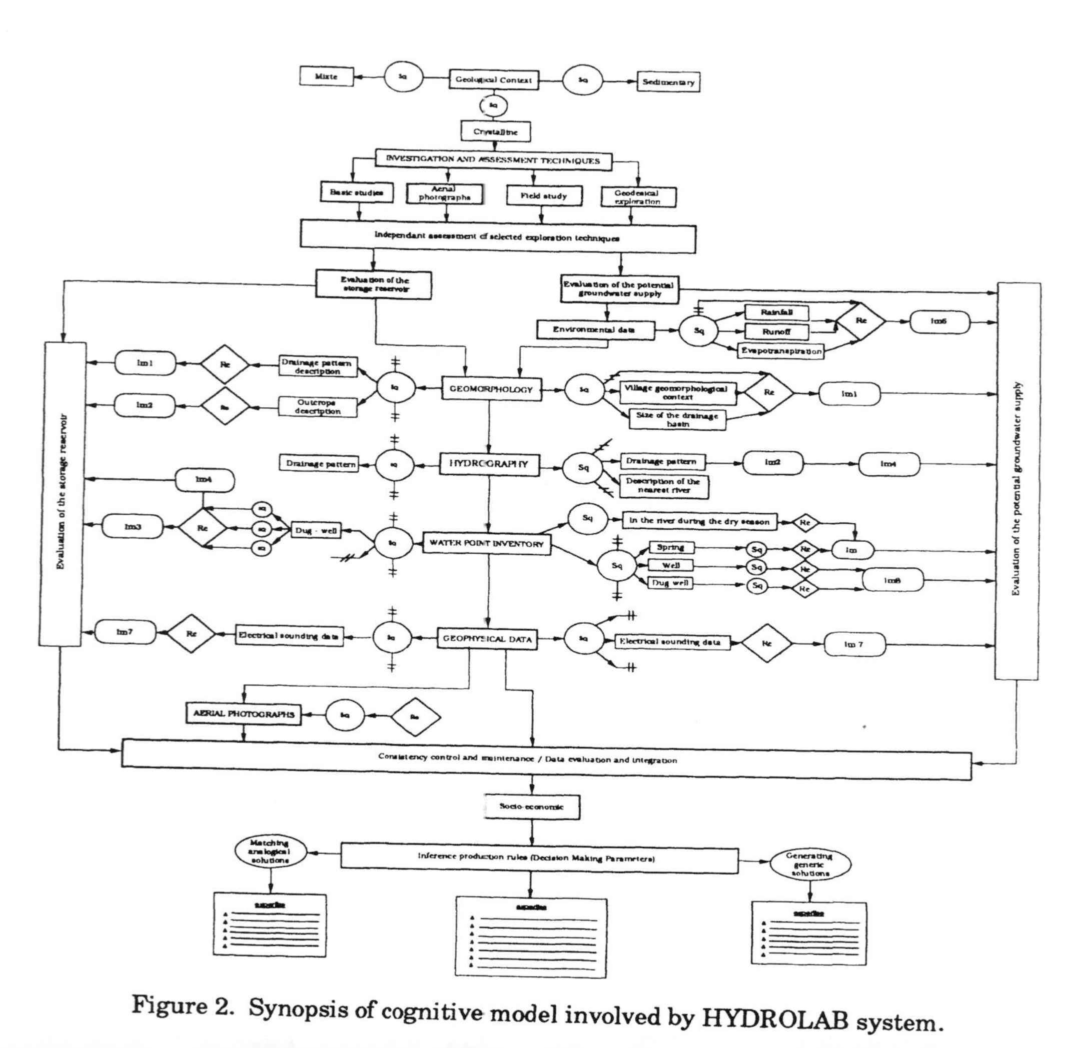

Figure 2. Synopsis of cognitive model involved by HYDROLAB system.

reasoning models involved and the functional organization of the various tasks to be carried out before a recommendation can be proposed by the system.

Before giving more insight to the relevant first-order decisional variables, considering the Basement Context modeled here, it is worthwhile to recall the conceptual model of the crystalline aquifer generally used by experts. In a crystalline environment, the storage and drainage functions coexist in each aquifer level, the weathered zone being essentially a storage layer and the bedrock being a drainage system.

The conceptual model of crystalline aquifer, which generally is accepted, is made of a semiconfining overburden (saprolite reservoir)–

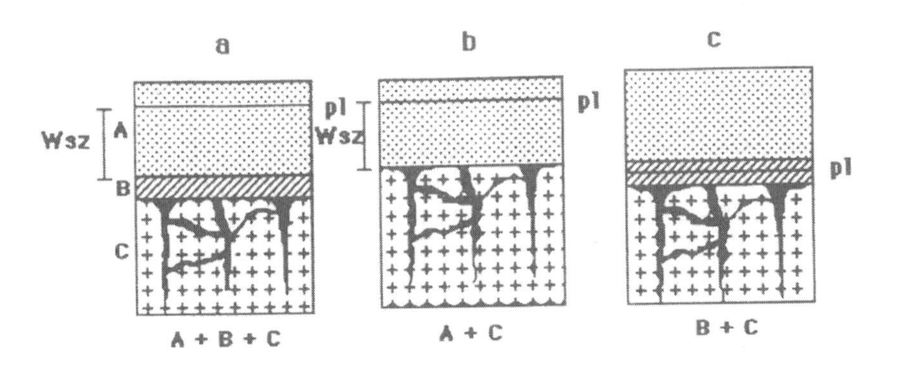

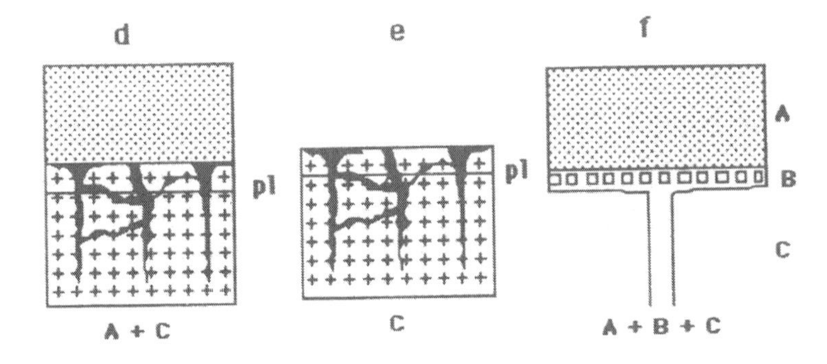

Figure 3. Conceptual model of aquifers systems in crystalline environment.

A: weathered zone (Ea) (i.e., storage function corresponding to a saprolite reservoir)
B: fissured zone (i.e., drainage function corresponding to the bedrock aquifer)
C: fault zone (i.e., drainage function corresponding to the bedrock aquifer)
D: two-layered groundwater reservoir
E: single-layered aquifer of saprolite, fissures or faults
F: equivalent model of the aquifer

this is mainly a storage reservoir, supplied from the surface—overhanging a fissure and fault confined aquifer (i.e., bedrock), draining the overburden and essentially having a drainage function. Although being not exhaustive, the aim of this scheme is to facilitate the reasoning. In this situation, it is considered that the aquifer system in the crystalline environment presents the structure of a two-layered (Fig. 3B, 3C, 3D) or even multilayered groundwater reservoir (Fig. 3A, 3F). However, single-layered aquifers of saprolite, fissures or faults (Fig. 3E) may exist separately (Bernardi and others, 1989).

The position of the piezometric level: This variable has proven to be a reliable first-order decisive element. Referring to the previous conceptual model, the piezometric level can be located either in a porous zone (weathered zone), in the fissured zone or in the fractured zone. In North Cameroon, we observed that the distribution of the piezometric level obtained from 527 observations shows that in 88, 6% of the situations, in a crystalline environment, the piezometric level is located in the weathered zone, corresponding to the saprolite reservoir (Fig. 4) and (Fig. 2, Im. 3).

There is a strong positive correlation between the presence of a saturated saprolite environment and the yield values obtained in the water catchment facilities. In terms of rules of expertise, the position of the piezometric level in the weathered rocks is a favorable element in the search for groundwater, and conversely its absence introduces a high percentage of risk (Re. 2). The position of the watertable and its associated variations during the year (especially during the dry season) can be recorded thanks to the observation of the pre-existing traditional wells dug by the villagers. These data are provided to the expert system during the water point inventory phase (§ 3.2).

Extent and thickness of the weathered layer: the in-situ weathered overburden thickness and the connections existing with the hydrodynamic characteristics of the aquifer, have been expressed thanks to a statistical approach by (Detay, 1987), using thousands of drillings carried out in the basement context. Probabilistic functions have been deduced showing the connections between the size of the weathered zone, the hydrodynamic properties of the aquifer measured by such characteristics as the yield or the specific yield, and the percentage of chances of obtaining the infra-yield (Fig. 5, Im. 4). The indications given by such graphics are connected to the decision-making parameters used by the expert system. We must add that the assessment of the thickness of the weathered layer involves a lot of basic data such as the depth of the

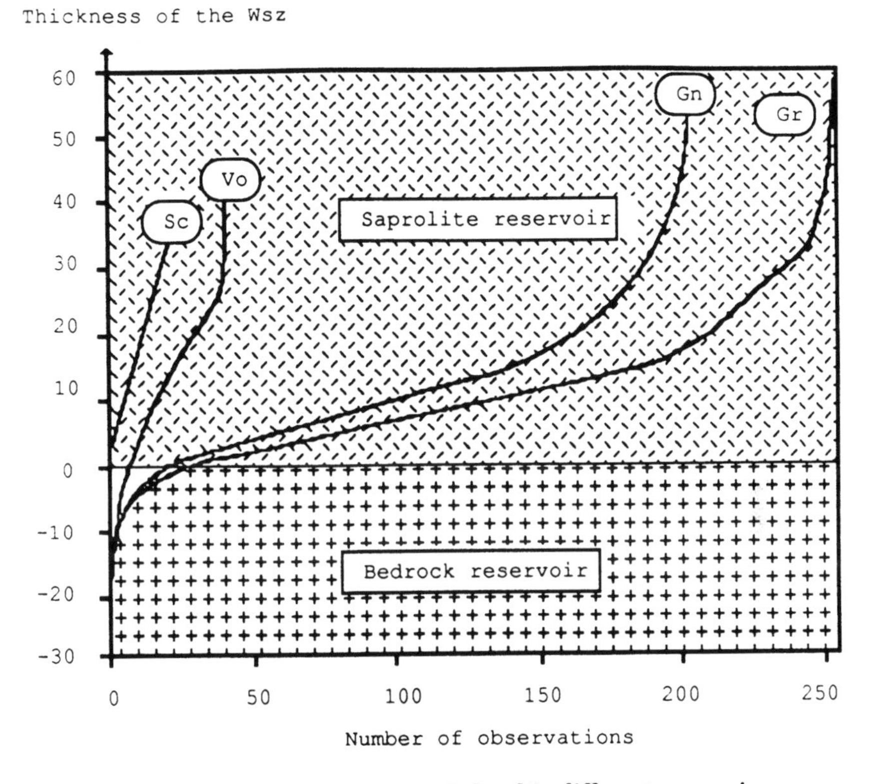

Figure 4. Position of piezometric level in different reservoirs.
Gn = gneisses (205 observations); Gr = granites (256 observations);
Vo = volcanic (43 observations); Sc = schists (23 observations).

traditional wells (villagers dig up to the compact and tough rock layer), the depth of the electrically resistant layer, on field observations, etc.

A thorough study allows us to study the influence of Ea on the yield (Q), the specific yield (Qs), and the probability of obtaining the infra yield (iQ), (Fig. 5). The yield expressed in m3/h, has been studied from pumping tests carried out on 540 operating drillings. The curve Q(Ea) shows the positive relation between the increase of Ea and the yield. The curve Q(Ea) is interpolated by the irregular rational function (1):

$$Q(E_a) = \left[\frac{1.07\,E_a + 0.2}{0.325\,E_a + 1.3} \right] \tag{1}$$

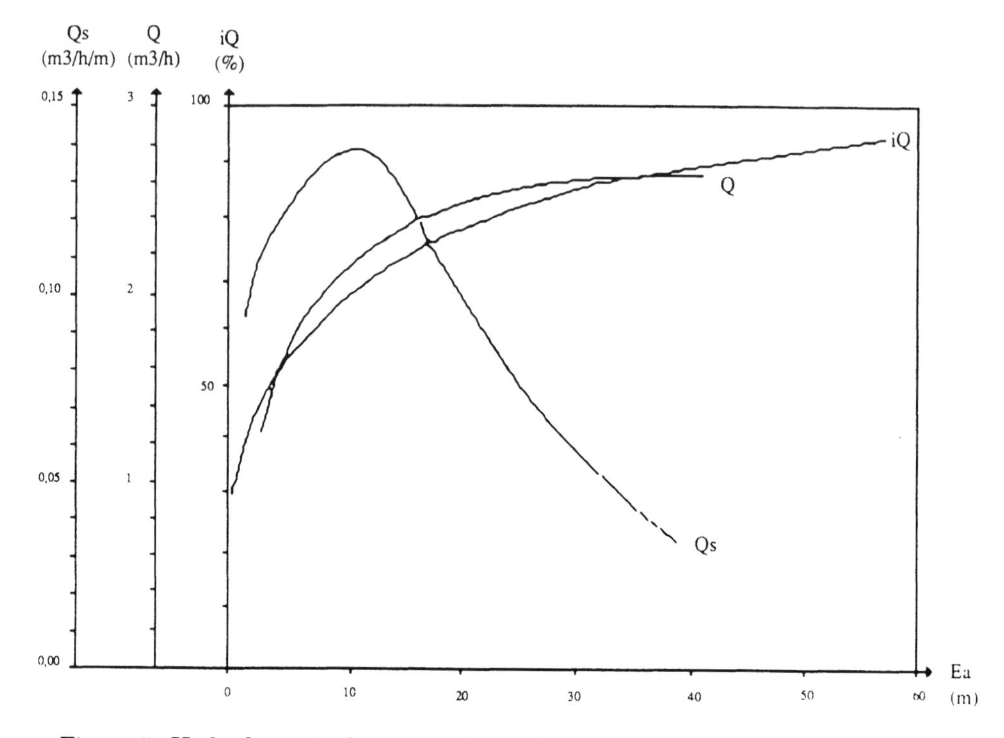

Figure 5. Hydrodynamic characteristics and success rate according to thickness of saprolite layer (Wl represent thickness of weathered layer, iQ for success rate (i.e., infrayield of 0, 75 m3/h), Q for yield in m3/h, and Qs for specific yield in m3/h/m).

The curve Qs(Ea), determined from 176 measurements, shows a linear increase in the first 13 meters before dropping (critical threshold) to tend to reach values less than 1.10-1 m3/h/m beyond 35 meters in the saprolite zone. Qs(Ea) is interpolated by the regular rational function (2):

$$Q_s(E_a) = \left| \frac{0.76\,E_a + 0.34}{0.0075\,E_a^3 - 0.145\,E_a^2 + 2.68\,E_a + 2.34} \right| \tag{2}$$

The probability, expressed in percentage, to obtain a minimum yield (fixed here at 0,5 m3/h) has been determined from a series of 715 drillings. The curve iQ(Ea) is interpolated by the irregular rational function (3):

$$iQ(E_a) = \left[\frac{0.7\,E_a + 0.9}{0.07\,E_a + 0.055} \right] \tag{3}$$

The functions Q(Ea), Qs(Ea), and iQ(Ea) apply for positive or less than 60 meters values of Ea (limits of the sampling). These results, accounting for the interpretative model Im. 4, can fit into the state-of-the-art as follows: the flows in a fractured zone are directed by fissuring, they generally remain laminar and comply with the Darcy's law. The yield coefficient being proportionate to the cube of the fissure width, the progressive blocking of the fissures in relation to the overburden increase, mainly clayed, results in a stabilization of the yield, and the exponential decrease of the specific yield when Ea increases. The probabilistic curves proposed clearly show the critical thresholds: 30 m for Q(Ea), 13 m for Qs(Ea).

Influence of the thickness of the saturated weathered zone: Thanks to kriging, we have been able to represent geometrically (Fig. 6) the relation between the thickness of the saprolite reservoir, the importance of the saturated weathered zone and the yield (Im. 8). The thickness of the saturated weathered zone may be negative when the piezometric level is situated in the bedrock. The yield seems to be directly linked to the thickness of the saturated zone. This relation reflects the mechanism of the aquifer system.

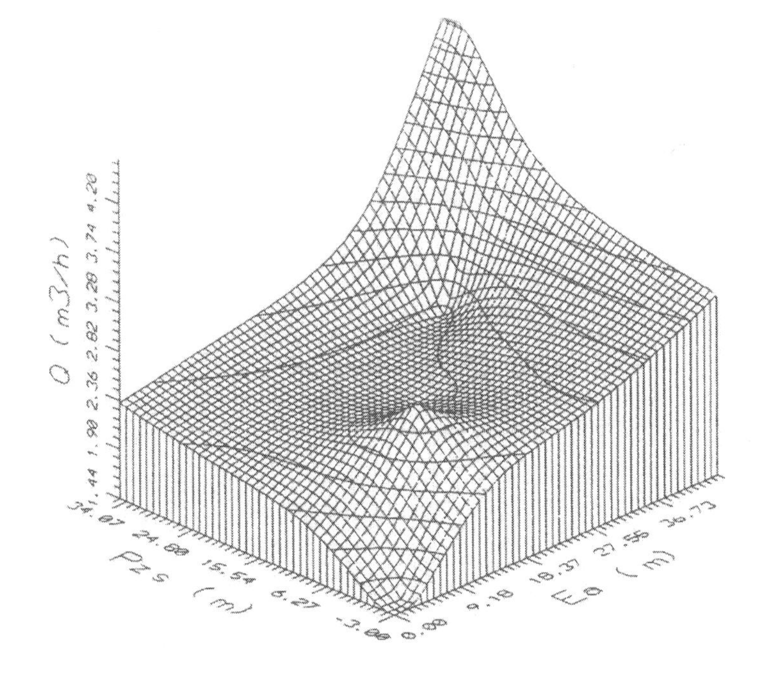

Figure 6. Influence of thickness of saturated weathered zone.

The water availability variable: in view of the size of the catchment basin, a balance can be computed for the distribution of the meteoric water (measured from the class of pluviometry the village is belonging to) between evapotranspiration (computed thanks to the Turc's formula for example), runoff (evaluated from measured yields at sample points of the drainage network), and effective infiltration (generally estimated as the unknown variable). This element which helps quantify the global amount of water available in the area, is not indicative of the water intake for a particular place as many other parameters should be considered including for example the properties of the sheet-type joint system or the relative development of the reservoirs (it will be done to assess second-order variables). But a careful examination shows that precise relationships exist between the infiltration capacity aforementioned and the probability to get a positive drilling and this computation gives straightoff a way to classify the area studied within a context that will be reused by any other decisional variables. Moreover, useful rules of thumb can be extracted from these observations and for example a practical corollary is that for the basement context studied here, an empirical link exists between the efficient rainfall and the hydrodynamic characteristics permitting to identify three domains where a minimum size of catchment basin is needed to get a positive drilling. These are respectively, 8 km^2, 5km^2, and 3km^2 for the domains of efficient rainfall of below 125 mm/year, between 125 and 500 mm/year, and more than 500 mm/year, respectively. The related rules are referred to as Re. 6 on Figure 2, and are connected to the interpretative model Im6 contributing to the evaluation of potential groundwater supply, one of the most important second-order decisional variables.

The drainage pattern variable: the size of the stream, its declivity, its profile, the shape of the bed and the morphology, are elements that permit an evaluation of groundwater recharge and discharge together with varying associated water levels, water quality, and pollution. This first-order variable is especially meaningful when the location is likely to concern underflow streams close to the village and when the aquifer is dependent mainly on the related marsh. In such a situation, the perenniality of the resources can be threatened during the dry season, and this variable is combined with other first-order variables such as the thickness of the weathered layer or the intensity of the sheet-type joint system, for example, to evaluate in a proper way the potentiality of the area. A complex arbitration between first-order parameters may be carried out, and the methods we used will be emphasized to illustrate the computation of second-order variables enabling generic reasoning functions (§ 4.6). Interpretative models (Im. 2) showing the connections between the drainage pattern, the average depth of the crystalline basement, and the percentage of failure in the drillings, are helpful is such situations.

The rock-type variable: the water-bearing characteristics of most crystalline rocks are controlled primarily by weathering and structure (Re. 4). Rock type alone usually is of secondary importance. Differences in well yields tend to reflect differences in degree of weathering or fracturing rather than inherent differences of mineralogy or fabric within the rocks (Re. 5). It should be noticed anyway, that the thickness of the weathered layer is significantly more developed in Western Africa, at the confidence threshold of 95%, on the granites than on the gneiss. This was established according to a statistical analysis we carried out to assess the significance of the difference of the means for two separate samples of these crystalline rock populations (respectively granites and gneiss). The only interesting issue is the potential implication of this variable on the development of the water-bearing reservoir. Rock type alone is not useful and so must be considered in a more general frame, including the amount and the properties of the weathering products generated. Special attention should be paid to the development of interstitial clays, apt to fill in the fracture network and to deteriorate the hydrogeological potential of the area.

The sheet-type joint system variable: It can be studied from aerial photographs. There may be a correlation between fracturing and the drainage pattern which indicates the potential water-bearing fracture system. Rules of expertise can be based, in hard rocks, on the evaluation of the density and extent of the system of cracks and fractures. Schematic searches can be made in the catchment area for major multikilometric fractures where the maximum volume of saprolite reservoir can be drained (Re. 3). For some situations, and especially when the weathered zone is under-developed, the only chance to get a positive drilling relies on the optimal use of the fracture network and on the search for favorable connections between subnetworks that join together. Nevertheless, one should be careful when confronted with an over-developed network of cracks, faults, and fractures, clearly discernible on the aerial pictures as it shows the thinness of the weathered layer, which is an unfavorable indication. Once more, intricate connections are displayed here between most of these first-order decisional variables, and their integration will be the result of an aggregating process described in the next section.

Longitudinal conductance variable: the research work carried out by our team since 1975 (Bernardi and Mouton, 1975; Bernardi and Detay, 1989) shows a relationship between the registered longitudinal conductances issued from the geoelectrical investigation (computed as the ratio between the expected thickness of the layer and its measured resistivity), the percentage of failure in the drillings and the hydrodynamic characteristics of the aquifer (Table 1). The definite total longitudinal conductance on the electrical soundings (from the position of the final ascending branch of the diagrams) corresponds to $\xi\iota/\chi\iota$ in which $\xi\iota$ represents the thickness of the different layers that compose the complex, and $\chi\iota$ their resistivity. Table 1, established after 273 drillings, carried out in Togo, Benin, Central African Republic, Cameroon, and Burkina Faso, statistically indicates that relatively high yields correspond to the high values of the total longitudinal conductance and inversely, low yields correspond to the low values of longitudinal conductance. Rules of expertise and interpretative models can be deduced, and offer efficient and reliable support in the decisional process.

The obvious meaning here is to seek for both conductive and thick layers, that represent of course favorable objectives. For more detailed information refer to Bernardi and Detay (1989).

The geomorphological variable: the first stage in photointerpretation and in the field study is a phase of observation. Topography has been determined to be an important indication of well yield in certain regions, as drillings on flat uplands and in valleys tend to yield larger amounts of water than those on valley sides and sharp hill tops. Rules of thumb, as elementary as (Re. 1): "in hard rocks move away from crystalline domes, mountain crests and sides which are non-water bearing" are nevertheless useful to avoid obvious mistakes sometimes observed. The lack of water on or near the steeper slopes can be explained by the fact that erosion has removed much of the weathered and more permeable rocks and by the intensity of the runoff. Water levels also are farther below the surface, because groundwater drains to points of discharge in adjacent lowlands. This decisional variable helpful in itself (making it

Table 1. Relationship between yield, success rate and longitudinal conductance (Im. 7)

Conductance (mho)	Villages	Success rate (%)	Yield m3/h)
inf. to 0, 1	48	inf. to 25%	inf. to 1
0, 1 to 0, 2	61	78 %	2
0, 2 to 0, 5	88	98 %	2, 8
0, 5 to 2, 5	76	66 %	3, 5

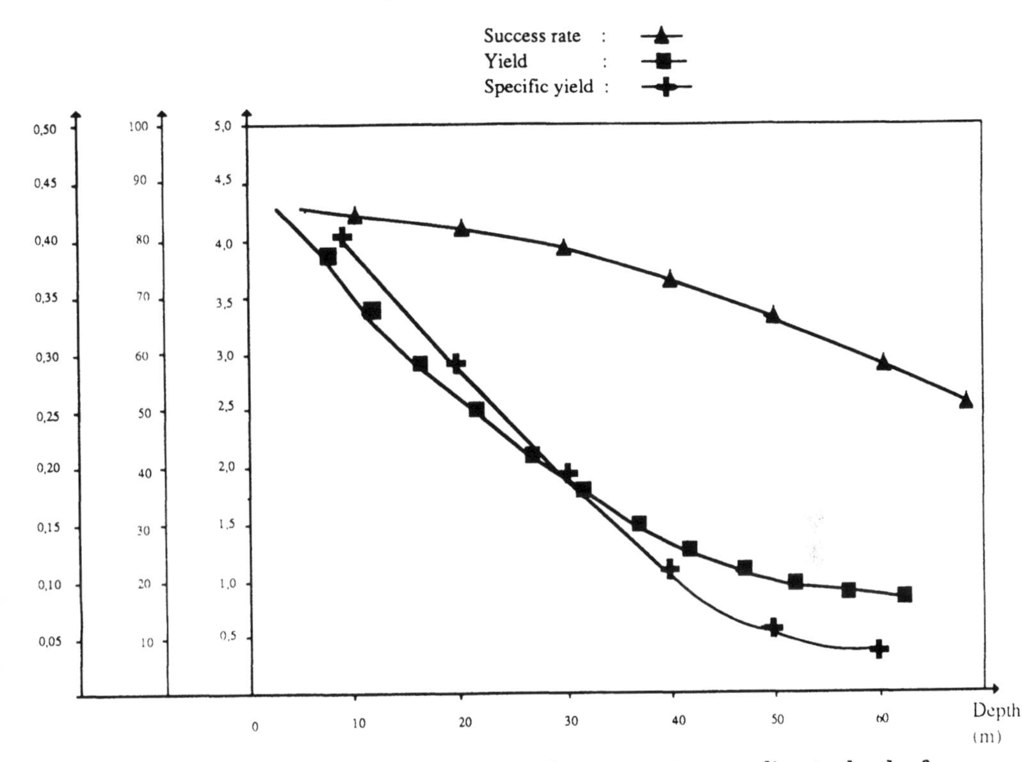

Figure 7. Hydrodynamic characteristics and success rate according to depth of drilling in basement complex.

possible to avoid incredible errors), is profitable fully when coupled with some of the previously described first-order variables (Im. 1).

The depth of borings - is there an optimal depth ?: this variable is the only one the operator can influence. To attempt to answer this question we worked on the depths of penetration into the crystalline basement itself (i.e., the total depth of the boring minus the size of the weathered zone). With 427 borings as a starting point, we calculated the depth of penetration into the crystalline basement by 5-meter segments of increasing depth, and we calculated the average yield in cubic meters per hour (m3/h) per segment (Fig. 7). An interpretative model has been deduced (Im. 5), showing the relationship between well yield per unit length of well penetration in the aquifer. It shows from 5 to 50 meters of drilling into the crystalline basement a negative relationship between the increase in depth and the increase in yield and from 50 to 60 meters a positive relationship. The water production per meter of well decreases rapidly with an increase in well depth. The optimum depth of water wells in crystalline rocks is determined mainly by economic factors unless the geological structure is known in detail.

Second-order Variables

Each of the first-order variables previously recognized, and some of the basic data themselves can be used to match the current context with reference situations stored from water-supply programs, and thus lead to a type of reasoning named analogical inferences. Nevertheless, to take full advantage of the expertise embedded within the cognitive model it is worthwhile to assess second-order variables which represent a compiling of the overall knowledge available and of the interdependencies observed between individual first-order variables. This integration leads to two essential second-order variables, which aims at analyzing the water intake and the potentialities of the water-bearing layer considered as a target reservoir. The way in which this integration is achieved relies on a formal framework entitled multiple objective decision-making theory, tailored to handle the location problem, and permitting the aggregation of multiple criteria leading to the aforementioned two main second-order variables. Formally, a criterion aims to summarize, thanks to a function, the assessment of an action on various consequences. A criterion g is a function providing real values, defined on the set A of the possible actions so as to compare the result of two separate actions a and b, thanks to their functional result g(a) and g(b) (i.e., two real values). The criterion g leads to assert the following statement:

$$g(b) \geq g(a) \Rightarrow b \; Sg \; a \qquad (4)$$

where Sg is a binary relation having a semantic such as "at least as good as", concerned with the consequences mastered by the g criterion. Referring to the meaning of the previous definition of the relation Sg, b Sg a can cover situations ranging from indifference (b Ig a) to strict preference (b Pg a). Using the model of the true criterion it is possible to distinguish these situations stating that:

$$g(b) = g(a) \Leftrightarrow b \; Ig \; a \qquad (5)$$
$$g(b) > g(a) \Leftrightarrow b \; Pg \; a \qquad (6)$$

According to the strong assumptions made when defining both the criterion (i.e., here the g function) and the consequences of the actions, it is relevant to consider two thresholds making it possible to define the pseudo-criterion model. In such a situation we can replace the previous definitions by:

$$g(b) = g(a) <=> b \; Ig \; a \; si \; g(b) - g(a) \leq qg(g(a)) \tag{7}$$
$$g(b) > g(a) <=> b \; Pg \; a \; si \; g(b) - g(a) > pg(g(a)) \tag{8}$$

The situation not covered by these formulae:

$$qg(g(a)) < g(b) - g(a) < pg(g(a)) \tag{9}$$

corresponds to the shallow preference Qg defined as an undetermined situation ranging from indifference to strict preference, where the functions qg and pg are referred to as the indifference or preference threshold.

For a criterion to be useful it is necessary to define in an accurate way the exact process making it possible to associate for each potential action a, the real number g(a) characterizing the performance of this action according to the selected criterion. To illustrate the approach, let us consider the situation where the criterion g aims to handle only one consequence, a measure of which is τ. Usually, this parameter is not exactly known; an appraisal is given by its probability distribution, where [pa(g), pa(g)dτ] is the probability range when the parameter is enclosed between τ and τ+dτ. It is possible to compare various actions according to their mean values of τ, expressed for τ enclosed between m and M by:

$$g(a) = \int_m^M \tau \, p_a(g(\tau)) \, d\tau \tag{10}$$

Of course, this approach can be unsatisfactory considering the observed standard deviations. An action a can be preferred to an intrinsically "better" action b (i.e., g(b) > g(a)) if it is characterized by a smaller standard deviation (i.e., σ(pa(g)) < σ(pb(g))). Thus in this situation action b leads to a better value for the consequences (i.e., according to the probability measure), but with a greater uncertainty. To handle this difficulty we use a monotone transform u(τ), characterizing the semantic attached to the measure of τ:

$$g(a) = \int_m^M u(\tau) \, p_a(g(\tau)) \, d(\tau) \tag{11}$$

To properly use such type of criteria it is necessary to determine correctly the function u(τ) and to characterize the probability law in a realistic way.

Of course, an isolated criterion is in itself not significant, and to achieve a relevant assessment of second-order variables it is necessary to develop a more detailed model based on multicriteria analysis. Let us consider now a family of criteria, each one of them concerned only with an homogeneous (set of) consequence(s). The result is to model second-order variables as more complex criteria coming from the aggregation of some of the n individual criteria of the family F. The problem now is to be able to discriminate actions taking into account the entire set of criteria applying to them, thus achieving the performance aggregation. Let us define the relation S for the set of actions A, such as a S b stems for the action a is at least as favorable as the action b. The first possible reasoning to achieve aggregation is to consider the set of concordant criteria C(a S b) with the proposition a S b.

This approach relies only on the ordinal properties of the assessments made using the performance table built up given the criterion results. But one can notice that some conflicting criterion(s) can be so important (i.e., $g(b) >> g(a)$) that even if a majority of concordant observations was respected to assess the reliability of the proposition, some doubt can exist and the proposition can be rejected as a definite threshold is stated. Obviously this observation leads to the definition of substitution rates between the criteria to establish the way in which an increase of value x on the conflicting criterion i is balanced with an increase of y on the concordant criterion j.

Let us consider the well-site selection, the situation where we aggregate in two synthetic criteria (i.e., water-supply favorability and water load-bearing structure capabilities), the overall set of primary criteria (i.e., first-order variables). An aggregation function V is defined, stating:

$$g(a) = V(g1(a), g2(a), \dots, gn(a)) \tag{12}$$

Once this formulation is asserted, it is possible to compare any action in terms of indifference, strict preference (or shallow preference if discrimination thresholds were introduced), and to proceed to a prescription (i.e., here we evaluate the two main parameters on which the decision is based) according to the alpha problematic described in Roy and Bouyssou (1988) (i.e., selection of a restricted set of satisfying actions or satisfecums). The aggregation function generally takes one of the following forms:

$$g(a) = \sum_{j=1}^{n} k_j g_j(a) \tag{13}$$

or

$$g(a) = \sum_{j=1}^{n} k_j v_j (g_j(a)) \tag{14}$$

where the kj are positive coefficients and the vj are monotone and strictly growing functions. This approach requires that there remains no incomparability between the criteria and that substitution rates could be defined accurately. Let us state that the substitution rate between the criterion gj and gh at the point:

$$g^0 = \left(g_1^0, g_2^0, \ldots, g_n^0 \right) \tag{15}$$

of the space performance is the variation on the criterion gj making it possible to balance a variation of reference on the criterion gh. If we assess a variation on gh of one unit, the substitution rate is the number such that the action characterized by the performances at point g0 is not different than the action having the performance such as:

$$\left(g_1^0, g_2^0, \ldots, g_{j-1}^0, g_j^0 + r_{jh}(g), g_{j+1}^0, \ldots, g_{h-1}^0, g_h^0 + \delta g_h^0, g_{h+1}^0, \ldots, g_n^0 \right) \tag{16}$$

thus leading to express the substitution rate in an additive model such as:

$$r_{jh}(g) = g_h(a) - g_j(a) \tag{17}$$

or if some derivative conditions are satisfied, and the variation on the criterion is such that gh->0:

$$r_{jh}(g) = \frac{\delta V(g)}{\delta(g_j)} - \frac{\delta V(g)}{\delta(g_h)} \tag{18}$$

According to Equations (13) and (14) it is possible to compute the values of the second-order decisional variables, thus involving the substitution rates defined for the various criteria as stated by Equations (17) and (18). The predicates water-intake-favorability (V1) and water–bearing–favorability (V2) are in charge of this task and the variables V1 and V2 returned by the activation of the relevant PROLOG clauses make it possible to assess the performance reached by the current situation for these two decisive and aggregating variables.

Decision-making Parameters

Proper well location, however, is but one of a number of problems facing the prospective well owner. Poor water quality, biological contamination, future lowering of water levels in wells, and improper completion methods are some of the many problems dealt with frequently by hydrogeologists. To make relevant recommendations, the hydrogeologist uses a decision-making process. In order to make the expert system operational we undertook the task of quantifying decision-making-parameters (Dmp.).

We defined:

- First-order parameters, which help in the decision to set up well location studies. Figure 5 indicates the probabilist evolutionary tendencies of the hydrodynamic characteristics according to the thickness of the saprolite reservoir. The utilization of these graphs makes it possible to do the following:

* Dmp. 1.a: evaluate the percentage of chances of obtaining the infrayield. It measures the boring risk as a function of the thickness of the saprolite reservoir;
* Dmp. 1.b: evaluate the average potential yield limit of the future wells;
* Dmp. 1.c: evaluate the average potential specific yield limit of the future wells.

- Second-order parameters, which help in the decision-making process when carrying out the drilling. Depth and cost are second-order decision-making parameters. They intervene not only from an economic point of view, but also are decisive in the potential failure or success of the drilling process. Figure 6 permits to do the following:

* Dmp. 2.a: evaluate the chances of obtaining the infrayield. It measures the boring risk as a function of the depth of the drilling;
* Dmp. 2.b: decide to what extent it would be opportune to continue the boring or whether it would be better to stop drilling, taking into account the finance available for the project;

* Dmp. 2.c: evaluate at each moment by scales of boring depth the average yield and the average specific yield and their evolutionary tendencies.

- Third-order parameters, which help in the decision-making process to design the well equipment and to use properly completion methods.

Following the project requirements and the water needs, these decision-making parameters help the user to take the right decision.

Task Analysis

As a brief summary to this cognitive analysis, first-order decisional variables and their associated interpretative models help to define a basic assessment of the context allowing a low-level matching with characteristic situations according to metonymy, second-order decisional variables lead to a higher level diagnosis based on the recognition of generic models using synechdoche, and decision-making parameters occur as operational recommendations handling the hydrodynamic characteristics of the wells and the economic factors.

This cognitive model represents the fundamental framework used to develop the HYDROLAB consultant, and the system's architecture reproduces at the software level these cognitive components and the links observed.

THE EXPERT-SYSTEM APPROACH

Introduction

Expert systems come from applied research in artificial intelligence. Computer-aided software aims to help in the decision-making process, and are designed to solve complex problems while trying to imitate human reasoning. Unlike mathematical or stochastic models, which operate on precise numerical data, expert systems mainly use a symbolic representation of the problem, known as a knowledge base and an associated chain of reasoning called strategy of inference. Nevertheless, coupled expert systems may solicitate the services of numerical models to gather and compute specific data.

To make the presentation of the HYDROLAB expert system understandable we refer where possible to the concepts described in section 3 and summarized in Figures 1 and 2, such as the Re., the Im., and the Dmp. previously described. It has not been possible to overview the entire HYDROLAB process which is about 5500 lines of PROLOG language (Colmerauer, 1983; Clocksin and Mellish, 1984; Warren, 1977; Borland, 1986). We have to refer to other Re. or specific questions Sq. which we have not been able to describe in this paper. We apologize for this, for more information please refer to authors' references.

System Overview

From an operational viewpoint, the start-up and the use of the system is extremely simple. The user is encouraged to volunteer information in natural language with HYDROLAB in order to provide various parameters that correspond to the data gathered for the given environment (Fig. 8).

From a functional viewpoint, the system identifies the hydrogeological context and evaluates the nature of the risk in using the described Im. to recognize different sorts of solutions. Depending on the significance of the risk it may reconsider its approach and ask for additional information on some point or other it considers important, before making a diagnosis.

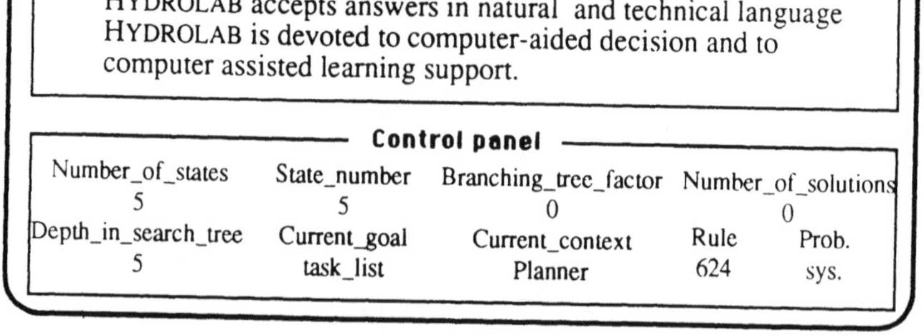

Interactive window of the Expert System

My name is HYDROLAB - I am a hydrogeologist consultant
My domain : groundwater location strategy
Would you like to use the automatic demonstration mode ?

Help window and/or current reasoning window

HYDROLAB accepts answers in natural and technical language
HYDROLAB is devoted to computer-aided decision and to computer assisted learning support.

Control panel

Number_of_states	State_number	Branching_tree_factor	Number_of_solutions	
5	5	0	0	
Depth_in_search_tree	Current_goal	Current_context	Rule	Prob.
5	task_list	Planner	624	sys.

Figure 8. HYDROLAB startup screen.

```
run :-
initializations,
/* ... */
planer,
!,
/* ... */
fundamental_inferences(State_depth),
/* ... */
conclusions,
shutdown,
browse_results.
```

Figure 9. First clause of HYDROLAB expert system.

From an interactive viewpoint, thanks to the system windowing it is possible for the user to understand the objectives followed by the software at any time, and to inquire about the implications of questions that have been asked. These functions confer to HYDROLAB computer-assisted learning (CAL) capabilities, moreover the system provides help facilities adapted to the type of request emitted by the operator.

Two types of solutions can be given by HYDROLAB: analogical ones are related to referential databases storing descriptions of local hydrogeological knowledge (i.e., case studies), and generic solutions provide a model of favorability for a given problem, based on abstract hydrogeological concepts (i.e., water intake, water bearing, etc.). Finally, decision-making parameters give technological information, derived from the adjustment of statistical laws for the observed distributions for large water-supply programs (i.e., thousands of drillings), enabling the user to forecast some of the drilling properties, thus representing a decisive help to the difficult location problem.

Description and Start-up

The system is composed of three main modules, as shown in Figure 9, a planer providing a way to trigger the tasks recognized during the cognitive analysis, an inference module devoted to the solution recognition, and a set of utilities allowing the display of the results.

The initial PROLOG clause to be evaluated is "run" where the /* ... */ pattern stands for many suppressed other clauses (Fig. 9). We reproduce here the PROLOG code, where predicates are English statements (original code is French), and variables to be unified are concepts enhanced by upper-case letter (PROLOG convention). Initializations, among other functions, aim to verify that the location submitted is within the scope of the program, and reject any proposal not conforming to the system's target (i.e., North Cameroon).

Data-gathering Phase

The planer is an important component, its role being to recognize the context encountered and then to trigger the relevant set of actions. The first step is to discriminate the context, that is, is the problem submitted corresponding to a basement, sedimentary, or mixed environment?, then to ensure that it will never be possible to reconsider this statement, the repeat clause avoiding backtracking until that point. As we emphasized in the cognitive analysis, reasoning is so exclusive for such different contexts that the system has to know soon if the cognitive model developed in this paper is relevant for the case submitted (i.e., basement context), or if it needs to apply a regionalized reasoning based on the geographical distribution of productive sedimentary layers, previously classified for each country considered (topics not considered in this article). Finally, according to this context selection, the software activates the control for the supposed context (Fig. 10).

The control clause is only a way to select a plan depending upon the filter Assessed_context (Fig. 11). The proper clause is selected among the three possibilities, the plan is retrieved (stored after its elaboration within the dynamic PROLOG database), and then scheduled. The slash (i.e. / or !) indicates that once the right clause is selected (and a plan is selected), the system will never reconsider other pending selections (i.e. remaining clauses).

```
planer :-
  /* ... */
  discriminate_context(Assessed_context),
  /* ... */
  repeat,
  retract_backtrack,
  control(Assessed_context), ! .
```

Figure 10. Planer discriminate probable context and schedules relevant set of tasks.

```
control(basement) :-
  basement_task_agenda(List_of_task),
  scheduler(List_of_task), ! .
control(sedimentary) :-
  sedimentary_task_agenda(List_of_task),
  scheduler(List_of_task), ! .
control(mixt) :-
  mixt_task_agenda(List_of_task),
  scheduler(List_of_task), ! .
```

Figure 11. Control clause permitting selection and triggering of adequate plan.

The List_of_task variable is a just a list of symbols, each of them being the name to be used to match with the filter of a task entry. This is an interesting property, in so far as the list is constructed dynamically and leads to a reactive behavior, using the data to pilot the control, the code itself being compiled and needing no runtime interpretation. The processing of that list, following the easier framework we developed, is straightforward, and just needs to satisfy a recursive call of the scheduler clause. For each head of the task list, the task predicate is invoked using the Task variable as a filter to retrieve selectively further suited clauses, and the processing ends when no symbol is left within the task agenda. In that situation we slash again, no pending selections having to be left here (Fig. 12).

For each task symbol extracted from the task agenda, a generic clause is used to process the task call. Three clauses are used to implement this function in a compact way, moreover allowing powerful control facilities, such as local or remove backtracking within the stack used to push and pop the tasks. The first clause is used to push the tasks in the stack, thus making it possible to go further in the selective and context-driven data collecting process, representing the first activity to be achieved (refer to Fig. 1). Variables such as Goal, Context, Rule, Proba, and Message are unified for the current task, then used to update

```
scheduler([]) :- ! .
scheduler([Task|Q]) :-
  task(Task),
  scheduler(Q), ! .
```

Figure 12. Scheduler clause allowing processing of task agenda.

For each task to be scheduled, once the Answer corresponding to Message is acquired, the system goes to the task_entry clause. If the triggering of the task is requested by the user (i.e., the frontchar of the tracking. Thus if the user wishes to return to a previous point in the past dialog (i.e., to change his mind for any given value or parameter for example), the task_entry predicate will fail before entering the task_entry clause, and will assert a backtrack fact within dynamic memory if a remote backtrack call was detected (i.e., operator will backtrack more than one step before), leading to pop from the stack as many task entries as necessary to reach the desired action. None of the two other clauses will be at that point eligible whatever happens, and the system will have to come back from at least one recursive call (i.e., thus popping the stack of the active tasks). Then either a backtrack fact has been introduced within the dynamic database, and the system clears the work previously done, then slashes (!) and fails until it reaches the right remote task and finally uses the third clause when the desired task is reached, or directly enter the last and third clause in a step by step backtrack to achieve a new recursive call for the proper task (Fig. 13).

```
task(Task_name) :-
  get_pannel(Task_name, Goal, Context, Rule, Proba),
  get_message(Task_name, Message),
  update_blackboard(Goal, Context, Rule, Proba),
  shiftwindow(2),
  print_optional_message(Task_name),
  change_activity(active, 2),
  task_entry_verification(Message, Answer, Task_name),
  change_activity(inactive, 2),
  task_entry(Task_name, Answer).
task(Task_name) :-
  backtrack(Symbol),
  Symbol <> Task_name,
  clear(Task_name), /* destructive backtrack */
  ! ,
  fail.
task(Task_name) :-
  not(generated_backtrack(Task_name)), ! ,
  clear(Task_name),
  retract_backtrack,
  task(Task_name).
```

Figure 13. Task generic entry providing powerful control facilities.

```
task_entry(Task_name, Answer) :-
  frontchar(Answer, 'y', _), ! ,
  sub_task(Task_name).
task_entry(Task_name, Answer) :-
  frontchar(Answer, 'n', _), ! ,
  negative_optional_action(Task_name).
task_entry(Task_name, Answer) :-
  backtract_solicitated(Answer),
  !,
  retract_backtract_generator,
  assertz(generated_backtrack(Task_name)),
  fail.
task_entry(_, _) :- ! .
```

Figure 14. Task_entry predicate based on natural language analysis to provide control functions.

the system interface control panel thanks to the update_blackboard clause. An optional message can be printed, then the system switches its activity level to trigger the control facilities providing access to back-Answer is 'y' for yes), the sub_task predicate is reached. In such a situation, it is of major importance to slash (!), in order to suppress any other backtracking possibility that could be generated if the sub_task predicate was to fail.

If not, two situations can arise: either the operator does not want to consider this action and HYDROLAB can decide to achieve some complementary optional actions depending on the current task, or the natural language analysis of the Answer shows that a backtrack solicitation has been requested (i.e., step-by-step or remove) and the software slashes (!) to suppress any other clause to be used, then assert a "generated_backtrack(Task_name)" fact within dynamic database, and finally fails so as to make the entire task predicate call unsuccessful (refer to explanations given for Fig. 14) In such an instance the processing will be reactivated at the task predicate level.

From that point, for each sub_task the required elementary data are gathered and asserted within the PROLOG dynamic database, thus making it possible to assess the values of the first-order decisional variables.

This process goes on until the blackboard is supposed to be filled. Then the first inferential process can start, leading for the best situations to the immediate recognition of analogical solutions (Fig. 15).

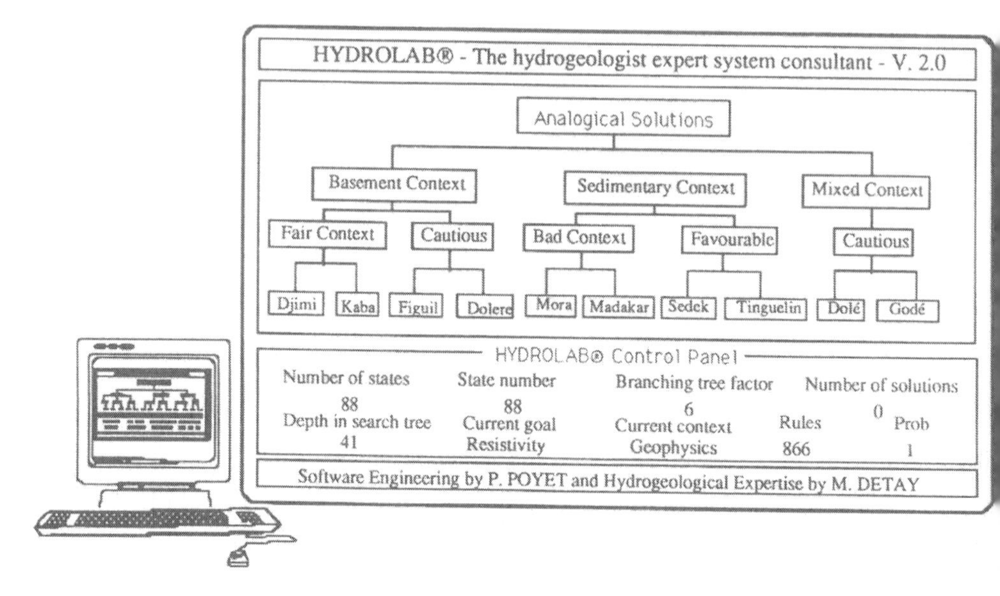

Figure 15. Example of HYDROLAB screen display during matching phase in analogical solution tree.

Inferences, Matching, Dynamic Planning

Referring to the section on "Description and Start-up"and Figure 9, one can see that the next step is to trigger the inference mechanisms both to produce the remaining reliable first-order decisional variables, and moreover to recognize analogical solutions using metonymy. This work is done thanks to the fundamental_inferences predicate, written so as to activate all the possible inferences.

Once an inference has been made, the predicate fails, then backtrack generates next inference, and the process repeats until the last one is reached. The predicate then is satisfied using the second clause, and the system slashes (!), (Fig. 16).

The "inferences" predicate is subdivided into four clauses. The first one is straightforward and directly matches with the analogical solu-

```
fundamental_inferences(State_depth),
  inferences(Prof_etat),
  fail.
fundamental_inferences(_) :- ! .
```

Figure 16. Triggering of inferential mechanisms.

```
inferences(State_depth) :-

      /* Matching with analogical solutions */

      /* set of fuzzy predicates enabling the fuzzy matching
       of the current situation with each database reference
       considered, thanks to first order decisional
       variables,
       leading for each to a global matching level */

      /* Solutions recognized are asserted within PROLOG
       dynamic memory database */
```

Figure 17. First clause of inferences predicate matches with analogical solutions.

tions stored within the databases of drillings to be considered as references (i.e. representative case studies). Side effects are represented mainly as screen update, and tree structure displays. This first clause is applied for all the known references (Fig. 17), until a solution can be discovered by fuzzy matching (Poyet and Detay, 1989c, 1989d), thus satisfying the inference predicate. In this situation, first-order decisional variables enable the system to recognize a typical pattern, and the software directly jumps to the conclusions, and expresses some forecasts according to the decision making parameters.

Otherwise, the second clause is used, once a new plan has been triggered to complement the basic data and the set of first-order decisional variables (i.e. this is the purpose of the third clause), to assess the favorability of the water intake (V1) and of the water-bearing layer (V2), before scheduling the generic inferences based on these second-order decisional variables.

The variables V1 and V2 are the result of a complex evaluation of the first-order decisional variables to be able, finally, to summarize the favorability thanks to two numerical parameters, allowing the discrimination of only one generic solution using synechdoche (Fig. 18).

```
inferences(State_depth) :-
      complementary_plan,
      fork(State_depth),
      water_intake_favorability(V1),
      water_bearing_favorability(V2),
      generic_inferences(State_depth, V1, V2).
```

Figure 18. Second clause aims to assess water intake and water-bearing favorability.

```
inferences(State_depth) :-
      fork(State_depth),
      not(solution(_, _)),
      not(complementary_plan),
      assess_complementary_plan(Set_of_actions),
      /* ... */
      scheduler(Set_of_actions),
      /* ... */
      inferences(New_state_depth).
```

Figure 19. Third clause collects proper complementary actions and enters recursion to trigger them.

If the third clause is reached, the system knows that it was impossible to recognize an analogical solution (i.e. first clause failed), and that a complementary gathering plan has not yet been done (i.e. second clause failed because the fact "complementary_plan" was not yet asserted within PROLOG dynamic memory).

So, HYDROLAB assesses a list of complementary actions to be accomplished, and returns a list of symbols, the Set_of_actions variable, used to enter a recursive call of the scheduler predicate, to ensure the triggering of the new plan (Fig. 19). Finally, the system makes a recursive call to the inferences predicate, thus restarting a complete phase of deduction, beginning with analogic models, then as a complementary plan was achieved with generic ones (i.e. clause 2).

When none of the previous clauses is applicable, the system must record a reasoning failure, this is the purpose of the fourth clause (Fig. 20).

Once solutions have been recognized and thus asserted in the PROLOG dynamic database either matched (i.e. analogic solutions) or built up from scratch by the system (i.e. generic ones), the fundamental task remains of reporting them to the user. As the recognition is based on a cumulative likelihood, an obvious pitfall would be to make a crude report which could incorporate nonsignificant observations in that few of

```
inferences(State_depth) :-
      fork(State_depth),
      not(solution(_, _)), ! ,
      /* ... */,
      assertz(solution(0, for_fun)).
```

Figure 20. Fourth clause registers reasoning failures.

```
rule(Number) :-
      db_open(ruledba, "rule.bin", in_file),
      bt_open(ruledba, "btree_rule", Btreesel),
      str_int(Key, Number),
      key_search(ruledba, Btreesel, Key, Ref),
      ref_term(ruledba, listestring, Ref, Message),
      /* ... */
      print_rule(Message),
      bt_close(ruledba, Btreesel),
      db_close(ruledba), ! .

rule(_) :-
      db_close(ruledba),
      fail.
```

Figure 21. First type of rule clause addressing database.

them could account for the main part of the overall variance observed. Thus, describing the work achieved and explaining the reasons for which a solution was selected is a real inference process and HYDROLAB uses the rule predicate to carry out this reasoning process, either for explaining for each first decisional variable taken into account factors leading to the analogic matching (Fig. 17) or, for second-order decisional variables, reasons leading to a particular generic case (Fig. 18). This rule predicate is built of hundreds of clauses which can be subdivided in two main groups. The first set is made of two clauses only, aiming to retrieve a database message corresponding to an expert prescription once the local corresponding reasoning has been done. The first one of these two clauses below uses the rule message number as a hashcode to generate a key making it possible to retrieve, thanks to the binary tree file, the location of the record within the database. Once retrieved, this message is reported to the user, and the two databases (i.e. data and binary tree selectors) are closed. The second clause is just used to close the database when the first attempt (i.e. first clause) fails (Fig. 21).

The second group of clauses achieves specific reasoning functions, according to thematic goals, to infer and explain for each first-order decisional variable considered the observed situation, the diagnosis made, and the related expert advice, to account for the analysis of second-order decisional variables and report the prescriptions related to the water intake and water-bearing structure favorability, and to assess the proper context to trigger numerical codes able to forecast hydrodynamic well properties according to the decision making parameters. To illustrate these concepts, we selected the rule numbered 749, aiming to provide a generic diagnosis related to the reservoir characteristics.

Figures 22, 23, and 24 illustrates, among many other hydrogeological patterns considered (i.e., many clauses have been suppressed), three possibilities deemed as representative of the type of analysis made to evaluate the reservoir properties.

The first clause concludes on the presence of a generalized aquifer, verifying the existence of a sedimentary context (i.e., the scope of the implemented program is far beyond the cognitive model presented here only concerned with the basement), checking for the absence of dry wells, assessing the thickness of the wet layer, etc. (i.e. many clauses were suppressed for intelligibility), evaluating the correctness of the water intake and reservoir favorability, finally reporting the result of its analysis (i.e., the rule (700) predicate call), and fails to permit the triggering of the lasting rules.

Figure 23, corresponds to the second clause for this rule numbered 749, and proposes to check for the existence of underflow streams enabling the water recharge, in the absence of a generalized aquifer. The reasoning is based mainly on the observation of dry wells (for a short time), of restricted water intake, and of proper characteristics for the given marsh.

```
rule(749) :-
        context(sedimentary, _, _, _), % any kind of sedimentary context
        no_dry_well(Vtest), Vtest=1,   % no dry well observed
        determine_water_table(Level),
        Useful_reservoir = Level + 10,
        wet_layer(Useful_reservoir, -9999, V1),
        V1 = 1,          % more than 10 meters of potential wet reservoir
        /* ... */                            % many clauses suppressed
        water_intake_quality(Note1),
        Note1 >= 12,                         % water intake is correct
        reservoir_quality(Note2),
        Note2 >= 12,                         % reservoir is correct
        rule(700),                   % we suppose a generalized aquifer.
        fail.
```

Figure 22. Second sort of rules concerned with diagnosis-like functions (generalized aquifer situation).

```
rule(749) :-
 dry_wells(0, 3, Vtest), Vtest = 1,
                       % wells are dry less than three months a year
  /* ... */            % many clauses suppressed
 water_intake_quality(Note1),
  Note1 <= 12,         % poor water intake
 presence_of_marsh(V1),
  V1 = 1,              % there is a marsh
 marsh_distance(-9999, 2, V2),
  V2 = 1,              % less than 2 kilometers far
 perenniality_marsh(10, -9999, V3),
  V3 = 1,              % water availability > 10 months
 rule(705),    % suspect underflow streams to produce water recharge
               % in the absence of a generalized aquifer
  fail.
```

Figure 23. Underflow streams could account for observations reported.

Going over many remaining clauses, Figure 24 is the last clause presented for this rule 749, and checks for the possible existence of a discontinuous fissured reservoir, mainly on the basis of the assessed water intake quality and on the existence of fractures and dry wells. Real reasoning is far more complex, but we tried to illustrate this second type of rule predicates with a concrete example.

This part of the real system is made of hundreds of rules, one set for each first-order decisional variable, one set again for each second-order decisional variables, and one for each decision-making parameter to trigger. HYDROLAB rules are written directly in PROLOG, leading to great efficiency as they are compiled by the PROLOG compiler in native code for the target machine.

```
rule(749) :-
 dry_wells(0, 3, V1),
  V1 = 1,              % wells are dry less than three months a year
  /* ... */            % many clauses suppressed
 water_intake_quality(Note1),
  Note1 <= 12,         % poor water intake
  fractures(_, _),     % fractures (whatever their type can be) exist
  rule(710),           % suppose a discontinuous fissured reservoir
  fail.

/* ... */              % MANY OTHER CLAUSES

rule(749) :- !.
```

Figure 24. Supposed existence of discontinuous fissured reservoir.

Results

The results provided by this research are twofold and must be evaluated in terms of experience gained from the development and the validation of the system and of benefits arising from the services offered by the software. Let us consider first the policies used to validate the environment and then the results obtained during the operational utilization. Automatic validation techniques applied to knowledge base integrity aim to guarantee the information consistency and to ensure the inference completeness. In that respect, research developments are mainly achieved in the formal context of propositional logics for monotonic bases (Ayel, Piparc, and Rousset, 1986; Le Beux and Fontaine, 1986; Pipard, 1988). Recent studies try to handle logics based on the first-order predicate calculus (Lalo, 1988; Rousset, 1987). Let us recall that a fact base is inconsistent if different values are encountered for the same property or attribute and that a rule base BR is potentially inconsistent if given a consistent fact base BF, the results provided by BF X BR are inconsistent; finally a rule base is declared incomplete if given a deductible attribute *A* there exists *v* one of the possible values *V(A)* for this attribute, such as for any initial consistent fact base BF, the couple (*A*, *v*) never belong to BF X BR. HYDROLAB relies on a first-order logic, its dynamic fact base is nonmonotonic (i.e., facts can be suppressed and the information flow is not monotonically growing) and facts handled by the system are intrinsically inconsistent. These properties stem from the development policies used and from the application domain particularities. Rules have been implemented thanks to the first-order logic of the PROLOG environment, facts can be suppressed as consequences of rules (nonmonotonic behavior), and different values can be assessed by various techniques for the same attribute (uprising fact inconsistencies). As a consequence, no automatic validation techniques can be used to guarantee the system coherency, and the correctness of the program can only be assessed thanks to robust software engineering policies based on incremental development and proper logistics. An unrealistic validation process would be to check the system for the overall set of not necessarily absolutely coherent but maximum fact bases *F*, that is, built of the set of undeductible facts supposed provided by the user, later saturating the fact base applying BF X BR and checking its coherency thanks to integrity constraints. This would not be a relevant process, observing that a subset of *F* can be extracted easily thanks to semantic constraints, considering the domain covered by the system. This subset of the maximum bases (defined as including the maximum set of undeductible coherent facts) can be used to check the system behavior for a representative set of hydrogeological situations, thus allowing the knowledge base coherency to be proven for a representative set of validating case studies. One should notice that initial bases are not necessarily absolutely coherent as some inconsistencies arise according to the logic of our application, and the aim is not to get rid of all inconsistencies but to provide mechanisms for the program. As was emphasized in the cognitive analysis section, considering for example the depth of the crystalline basement, various estimations lead to conflicting values, coming either from the depth of traditional wells, from an assessment of the thickness of the weathered layer, or from high-resistivity layers showed by geophysical prospecting. We provided the system with reasoning policies to handle such conflicting parameters, but no provision was made to handle wrong values which could be provided on purpose by the user to evaluate the system robustness. The strategy previously described makes it possible to handle conflicting rules thanks to the test bases but does not cover the completeness of the rule set.

A strict logistics was involved to achieve the prototype development, moreover taking advantage of robust software engineering policies leading to incremental and modularized implementation. The program requirements and prespecifications were established in collaboration with an expert. They were reused as a basis to establish a productive confrontation with different specialists coming from the many subdomains concerned by the decision-making process, thus enabling us to write detailed functional specifications leading to the first implementation. Taking advantage of the large rural water-supply programs we carried out in North Cameroon (1250 drillings), in the Central African Republic (350 drillings), and in Gabon (500 drillings) leading to the gathering of really numerous data, we designed the maximum coherent bases to be used as test modules for assessing the rule coherency.

Then the system was controlled as a practical exercise of a course we teach at the International Training Centre for Water Resources Management, at the Nancy National Geological Engineering School and at Nice University (France). The system then was presented to international congresses and modified to take into account suggested improvements.

Finally it was possible to install the first operational environment in different African technical establishments such as the rural engineering section of the Central African Republic and at Bangui University to collect a feedback from the effective daily practice of the software. It leads us to consider more industrial and in-situ problems, topics seldom covered by AI developments rarely going farther than the prototype stage in the majority of known projects (Yatabe and Fabbri, 1989).

Functional capabilities of the program are straightforward and lead to high-value services to handle the well-site selection. A working session with HYDROLAB is attractive for the users because of its ability to explain its own reasoning, to understand the questions asked in natural language, to explain its approach at any given time, and to undergo changes and acquire new data. But the real interest, of course, comes from the results provided by the system, which are expressed in terms of proximity to case studies (i.e., analogical solutions) and to generic descriptions of conceptual hydrogeological contexts (i.e., generic solutions). For each one of the solution types, the software explains its diagnosis, reports the observed semantic proximity and the rules involved to discover it, analyses the consequences of the decision-making parameters to forecast some of the hydrodynamic properties of the drilling once achieved. This clear macroscopic behavior of the expert system leads to a good operator understanding of the actions carried out by the software, probably accounting for the success encountered by the program for in-situ operations. Moreover its functional capabilities ensure high-success rates such as 80% in the situation of the drillings achieved under the program control in areas such as the North Cameroon.

CONCLUSION

The problem of village and rural water supply has to be handled at a continental scale, and must take into account the diversity of the encountered situations thanks to flexible and intelligent tools. The behavior of these systems must be mapped on the steps followed by specialists when handling various geological contexts. Taking advantage of an exhaustive modeling of the available location strategies for the North Cameroon area including either sedimentary or crystalline basement seepages, we developed a cognitive framework permitting us to cope with a more generic approach reliable for the basement context whatever the area.

We went through the identification of decisional variables of different orders and of decision-making parameters and we described the expert-system approach to model the process used by a hydrogeologist to carry out successfully drilling-sites studies. Taking advantage of the large amount of expertise formalized and of original knowledge modeling policies and software engineering techniques, it was possible to develop an extremely efficient operational software with a high performance level on low-cost microcomputers, a result which would not have been conceivable on large computers a few years ago.

We believe that in the near future the growth of this technology will make it possible to reduce considerably the cost of studies thanks to the training and the increased involvement of national experts in the development of their countries' water resources. Artificial intelligence therefore should be able to play a significant role in solving the water problem in Africa especially within the scope of programs for village water supply. We hope that artificial intelligence will furnish material of both practical and theoretical interest to water resources scientists and also to those involved in water-resource assessments and planning for water-resources management.

REFERENCES

Ayel, M., Pipard, E., and Rousset, M. C., 1986, Le contrôle de cohérence dans les bases de connaissances: Proc. des Journées Nationales sur l'Intelligence Artificielle, PRC-GRECO I.A., Cépadues Ed., November 20-21, p. 171-186.

Bernardi A., and M. Detay, 1988. Corrélations entre les paramètres géoélectriques et les caractéristiques hydrodynamiques des forages en zone de socle: Hydrogéologie, Revue BRGM, no. 4, p. 245-253.

Bernardi A., and J. Mouton, 1975, Recherche d'eau dans les formations cristallines et métamorphiques du socle Africain: Proc. of Intern. Congress of Hydrogeology, Porto Alègre, Brazil, p. 1–34.

Bourget, L., Camerlo, J., Fahy, J.C., and Vailleux, Y., 1980, Méthodologie de la recherche hydrogéologique en zone de socle cristallin: Bull. BRGM, Section III, no. 4, p.273-288.

Borland International, 1986, Turbo Prolog owner's handbook: Borland Publ., California, 221 p.

Colmerauer, A., 1983, Prolog in ten figures: Proc. of Eighth Intern. Joint Conference on Artificial Intelligence, Milan, Italy, p. 487-499.

Clocksin, W.F., and Mellish, C.S., 1984, Programming in PROLOG: Springer-Verlag, Berlin, 304 p.

Detay, M., 1987, Identification analytique et probabiliste des paramètres numériques et non-numériques et modélisation de la connaissance en hydrogéologie sub-sahélienne: Thèse de Doctorat d'Etat ès Sciences, Université de Nice, France, 456 p.

Detay, M., and Poyet, P., 1988a, HYDROEXPERT: Un système expert en hydrogéologie de terrain: Proc. HYDROPLAN Intern. Exhibition of Agricultural, Rural, Urban, and Industrial hydraulic Engineering, Marseille, France, p. 191-195.

Detay, M., and Poyet, P., 1988b. Environnement et géologie: application de l'intelligence artificielle en hydrogéologie: Géologues, Sciences de la Terre et Techniques Modernes, Revue Officielle de l'Union Française des Géologues, v. 2/3, no. 85-86, p. 41-47.

Detay, M., and Poyet, P., 1989a, Introduction aux méthodes modernes de maîtrise de l'eau: Hydrogéologie, Revue BRGM, 1990, no. 1, p. 3–25.

Detay, M., and Poyet, P., 1989b, La Place de l'informatique dans les géosciences: Géologues, Sciences de la Terre et Techniques Modernes, Revue Officielle de l'Union Française des Géologues, v. 4, no. 91, p. 37–49.

Detay, M., and Poyet P., 1990, HYDROLAB, an expert system for groundwater exploration and exploitation. Intern. Jour. Water Resources Development, v. 6, no. 3, p. 187-200.

Detay, M., Goulnik, Y., Casanova, R., Ballestracci, R., and Emsellem, Y., 1986,HYDROLAB an expert system for groundwater exploration in Africa: Bull. IWRA - Intern. Conf. on Water Resources Needs and Planning in Drought Prone Areas, Khartoum, Sudan, p. 1231-1236.

Detay, M., Poyet, P., Emsellem, Y., Bernardi, A., and Aubrac, G., 1989a, Influence du développement du réservoir capacitif d'altérites et de son état de saturation sur les caractéristiques hydrodynamiques des forages en zone de socle cristallin: Comptes Rendus de l'Académie des Sciences de Paris, t. 309, Série II, p. 429-436.

Lalo, A., 1988,TIBRE: Un système expert qui teste les incohérences dans les bases de règles: Proc. Eighth Intern. Workshop on Expert Systems and their Applications, v. 3, Avignon, France, p. 63-84.

Le Beux, P., and Fontaine, D., 1986, Un système d'acquisition de connaissances pour systèmes experts: Technique et Science Informatiques, v. 5, no.1, p. 12–34.

Pipard, E., 1988, Détection d'incohérences et d'incomplétudes dans les bases de règles: Le système INDE. Proc. Eighth Intern. Workshop on Expert Systems and their Applications, Avignon, France, v. 3, p. 15-33.

Poyet, P., 1986, Un système d'aide à la décision en prospection uranifère: méthodes de discrimination des anomalies géochimiques multiélémentaires significatives: Thèse de Doctorat d'Etat ès Sciences, Université de Nice - INRIA, France, 436 p.

Poyet, P., and Detay, M., 1988a, HYDROEXPERT: Aide à l'implantation d'ouvrages d'hydraulique villageoise: Proc. Eighth Int. Workshop on Expert Systems and their Applications, Avignon, v. 2, p. 397-410.

Poyet, P., and Detay, M., 1988b, L'avènement d'une génération de systèmes experts de terrain, *in* Demissie, M., and Stout, G. E., eds., Proc. of the Sahel Forum on the State-of-the-Art of Hydrology and Hydrogeology in the Arid and Semi Arid Areas of Africa: Ouagadougou, Burkina Faso, p. 567-577.

Poyet, P., and Detay, M., 1989a, Enjeux sociaux et industriels de l'intelligence artificielle en hydraulique villageoise: Proc. Première Convention Intelligence Artificielle, Hermes Publ., v. 2, Paris, p. 621-652.

Poyet, P., and Detay, M., 1989b, HYDROEXPERT Version V1.4, Un système expert d'aide à l'implantation de forages en hydraulique villageoise: Rapport de Recherche INRIA, no. 936, Déc. 1988, 43 p.

Poyet, P., and Detay, M., 1989c, HYDROLAB: Un système expert de poche en hydraulique villageoise: Technique et Science Informatiques, v. 8, no. 2, p. 157-167.

Poyet, P., and Detay, M., 1989d, HYDROLAB: A new generation of compact expert systems: Computers Geosciences, v. 15, no. 3, p. 255-267.

Roy, B., and Bouyssou, D., 1988, Aide à la décision, AFCET/INTERFACES, Mars 1988, no. 65, p. 4-13.

Rousset, M. C., 1987, Sur la validité des bases de connaissances: le système COVADIS: Proc. Seventh Intern. Workshop on Expert Systems and their Applications, Avignon, France, p. 269-282.

Warren, 1977, Implementing PROLOG, DAI Research Report v. 1 and v. 2, Univ. of Edinburgh.

Yatabe, S. M., and Fabbri, A. G., 1989, Putting AI to work in geosciences: Episodes, v. 12, no. 1, p. 10-17.

Hydrodat®: A PACKAGE FOR HYDROGEOCHEMICAL DATA MANAGEMENT

A. Minissale
C.N.R.- Centro Di Studio Per La Mineralogia E La Geochimica Dei Sedimenti, Florence, Italy

and

G.F. Buccianti
ECOSYSTEMS s.a.s., Florence, Italy

ABSTRACT

Hydrodat®: is a database system with reference to phreatic and running waters (more generally to liquid and condensate phases) and gas phases. This database system was written for personal computers running MS-DOS.

The three principal data files for physical and chemical characteristics which the program runs (one for phreatic waters, one for stream waters, and one for gases), the calculation routines—some from the literature, others specifically processed—and the graphic elaboration (available through "Grapher" and "Surfer" of Golden Software) are controlled by procedures using mainly dBase IV language of Asthon Tate.

The following points indicate the possible fields of investigations supported by the use of the database:

Use of Microcomputers in Geology, Edited by D.F. Merriam and H. Kürzl, Plenum Press, New York, 1992

(1) fluid's geochemistry;
(2) hydrochemical prospecting;
(3) geothermics;
(4) volcanology;
(5) environmental control;
(6) monitoring of industrial fluids.

The program requires a minimum configuration of 640 Kb of RAM memory and is supported entirely by window-menu and window-help instructions.

INTRODUCTION

The wide diffusion of personal computers, including laptops, in particular the new 32-bit 80386 INTEL equipped models, has made these machines convenient and economical.

Hydrodat®,whichwill be described and illustrated in the following paragraphs, is an example of relational database written using DB IV Asthon Tate language. It manages three data files of chemical analyses, graphic, and processing procedures.

The data handled by this program are physical parameters and chemical data of phreatic and stream waters—more generally liquid and condensate phases—and analyses of natural or industrial gas phases (Fig. 1). This database mainly responds to the needs of those concerned with fluid geochemistry, volcanic and geothermal fluids, springs and mineral waters, and stream waters. It also can be used for hydrochemical prospecting, chemical, and pollution monitoring in aqueducts, purification plants, etc. A further category to be processed by the program are isotopic data of fluid phases. In Figure 2 the window for groundwater data entry is shown as an example.

A more recent version of Hydrodat® has been developed for managing any type of environmental analyses data files. In this version, which runs all the graphic and statistic procedures of Hydrodat®, data entry windows and related units can be set up by the operator.

DATABASE STRUCTURE AND DATA-EXTRACTION CRITERIA

The input windows of the physical parameters and chemical species are divided into:

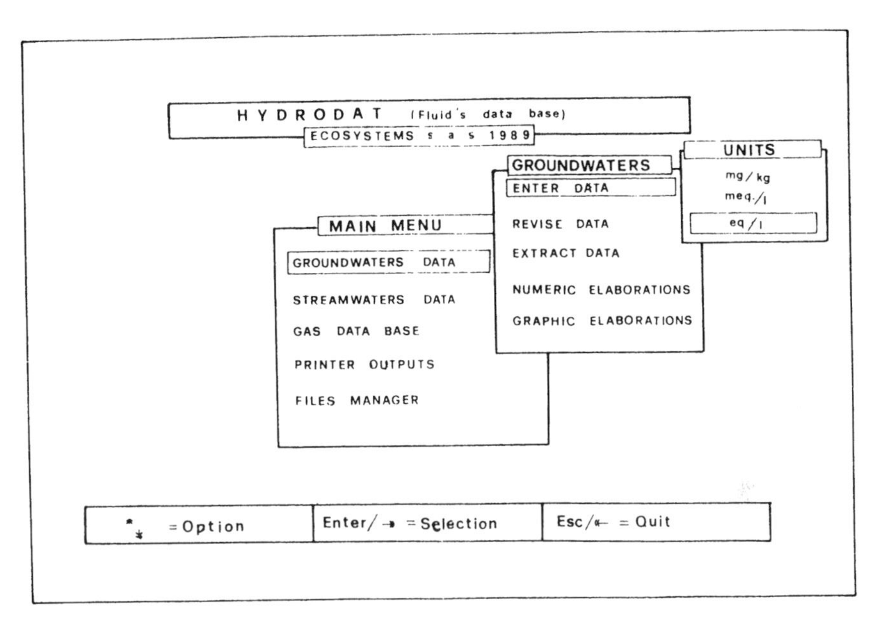

Figure 1. Starting window of program.

- GROUND WATERS -

NAME CODE DATE / / TYPE code AREA code ELEVATION

UNIT : ppm

FIELD DATA: T (°C) . pH . Cond. (S/m) . Alcalin. . lg pCO2 .

LABORATORY DATA: T (°C) pH . Cond. (S/m) . Alcalin. . T.D.S. ppm .

Ca++ . Mg++ . Na+ 3.33 +3 K+ . H2S . Cl- . SO4-- .

Li+ . Rb+ . Cs+ 9.77 -1 Ba++ . Sr++ . Zn++ . Pb++ .

As3+ . Fet . Mnt . Alt . Al+ . F- . Br- .

Hg+ SiO2 H3BO3 NH4+ gas/vap.(nl/kg) CO2 T.D.S. an. (meq/l) T.D.S. cat. (meq/l)

ISOTOPS

18O/w . 2H/w . 3H/w . 34S/SO4 . 18O/SO4 . 13C/HCO3 .

COORDINATES X: Y:

←↕→ = Move pointer | ^End = Confirm input | Esc = Previous screen

Figure 2. Window of data entry for groundwaters data-file.

(1) groundwaters (Fig. 2);
(2) stream waters;
(3) gases.

The two phreatic and running water input windows differ for the quality of the data processed. In particular the main polluting agents, both organic and inorganic, have been considered in the stream water record.

In the groundwater input window, pCO_2 and salinity (T.D.S.) of samples are calculated automatically at the time of data input and then are processable in their turn (Fig. 2).

In the gas record the program automatically reports if, for any given gas sample, there is a corresponding analysis in liquid phase. Moreover in regard to gases, the program automatically calculates the ratio between the $^3He/^4He$ and $^{40}Ar/^{36}Ar$ values in the sample and the corresponding atmospheric values.

For additional data input, the three main datafiles automatically fall into alphabetical order and selected input date order. This makes it easy to retrieve any sample for updating or deleting and, at the same time, it facilitates sequential extraction of data from the main data files and to process graphically thc extracted files in time.

A series of filtering procedures are provided by:

(1) name;
(2) code;
(3) input date;
(4) area code;
(5) morpholocical code (hot spring, acid water, boiling pool, etc.;
(6) chemical type (chloride, sulphide, or bicarbonate);
(7) elevation or depth range;
(8) temperature range;
(9) salinity range;
(10) time interval;
(11) any association of the previous 10 parameters.

They are released with the package, but every user can build easily his or her own procedure for extracting data to print or process. Descriptions for sampling areas, morphological types, etc., can be handled through accessory procedures that may or may not be transparent to the user.

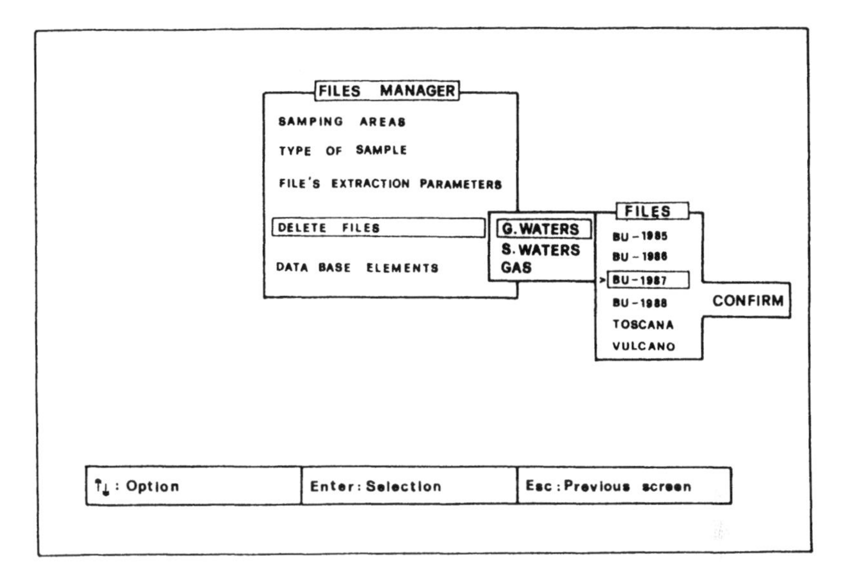

Figure 3. Input window of Files Manager which runs secondary extracted files as well as area's and morphological type's codes.

Once a series of parameters have been selected (e.g., the mentioned time), the program makes it possible to extract from the main data file one or more secondary datafiles for further processing. Similar extraction criteria are available for stream waters and the gas main databases, as well as for all databases, already mentioned for the recent generalized version of Hydrodat®.

As previously said, the procedure uses codes for sampling areas and morphological types, whose description can be handled through accessory tables. In Figure 3 the handling of the mentioned accessory codes (sampling areas and types of samples), and of the extracted secondary datafiles (visualization of extraction parameters, deletion of extracted files) is shown.

DATA PROCESSING

Both graphic and numeric processing can be done on extracted files. The following is an example for groundwater data:

(1) X-Y diagrams (including in time diagrams as in Fig. 4);
(2) triangular diagrams;
(3) histograms;
(4) Piper diagrams (only for groundwaters as in Fig 5).

For each of the mentioned diagrams, the drawing, the modification of data and symbols, as well as the monitored visualization and the output on plotter or graphic printer can be done through the Golden Software's "Grapher" program instructions.

In the cartesian and ternary diagrams, it is possible to select variables, both singular physical parameters and chemical species , as well as algebric combination of species and numeric quantities [e.g.: log(a), a/100, 1/a, (a+b)/c., etc.].

In time diagrams (Fig. 4) the program automatically selects the time scale on the basis of the age of samples.

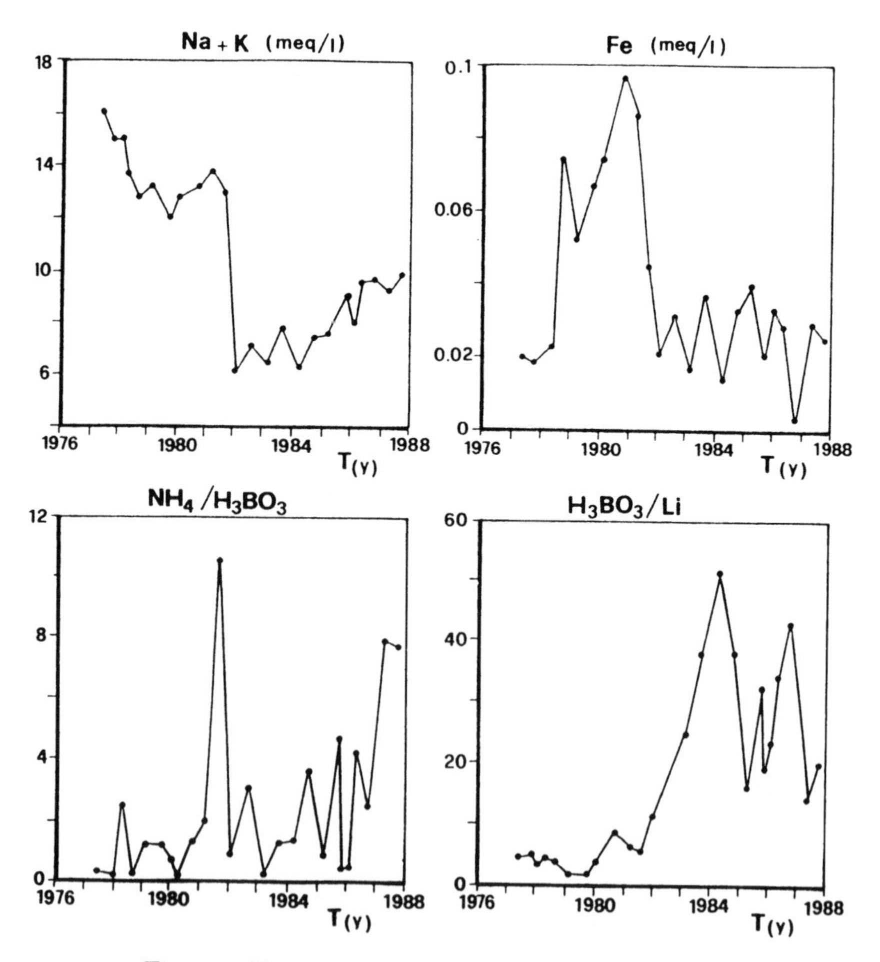

Figure 4. Example of in time variation X-Y diagrams.

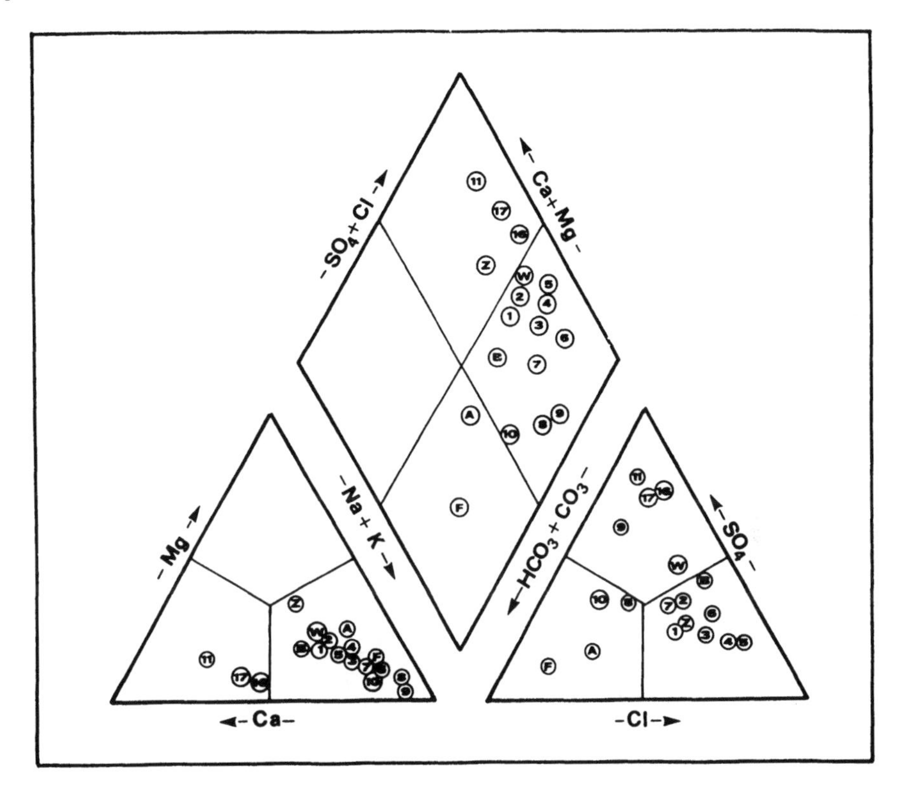

Figure 5. Example of Piper diagram output on plotter.

Numerical processing provides elementary statistics on single variables:

(1) the numbers of data;
(2) the minimum value;
(3) the maximum value;
(4) the sum total of values;
(5) the mean value;
(6) the variance
(7) the standard deviation;
(8) the % variation coefficient;

and multicomponent statistics:

Table 1. Correlation coefficients

** file = LARDEREL.WAT **** CORRELATION COEFFICIENTS ********* 25/05/1989 **

	t(°C)	HCO_3	H_2S	Cl	H_3BO_3	NH_4	g/v	H_2S_g	H_2	CH_4	N_2	NH_4/B
t(°C)	1.000	-.121	-.124	-.234	-.169	-.177	-.095	-.137	0.651	-.134	-.129	-.197
HCO_3		1.000	-.111	-.069	-.093	-.136	-.103	-.029	-.122	0.814	-.127	-.084
H_2S			1.000	-.155	-.049	-.154	-.134	-.105	-.119	-.109	0.752	-.127
Cl				1.000	-.143	-.071	-.163	-.154	-.227	-.040	-.132	0.550
H_3BO_3					1.000	-.127	-.046	-.107	-.178	-.105	-.054	-.129
NH_4						1.000	-.105	-.078	-.199	-.140	-.158	-.111
g/v							1.000	-.095	-.122	-.116	-.150	-.127
H_2S_g								1.000	-.149	-.026	-.133	-.142
H_2									1.000	-.103	-.087	-.209
CH_4										1.000	-.121	-.063
N_2											1.000	-.152
NH_4/B												1.000

Example on printer output of multivariate analysis (correlation coefficients) from secondary extracted file taken from the main groundwater data-dase.

(1) multiple correlation coefficients calculation and display (Table 1);
(2) cluster analysis with dendrogram display;
(3) factor analysis.

Single programs taken from the literature or specifically developed, allow the following calculations:

(1) ionic strength of solution (Table 2);
(2) activity coefficients and the activities of all species in solution (Table 2)
(3) solubility product (K_a) of the solution for calcite, anhydrite and fluorite (saturation condition for the solution in these mineral species (Table 3);
(4) partial pressure of CO_2 in solution;
(5) temperatures calculated with the main chemical geothermometers in liquid phase for geothermal purposes (Table 4).

Graphic processing provides presentations of parameters in areal distribution maps. For this purpose two numeric fields of relative coordinates (x,y) are available in each single input window (Fig. 2). By using these coordinates it is possible to generate—by the commercially

Table 2. Activity of chemical species in solution

** file = APUANE.WAT ********** A C T I V I T Y *********** 25/05/1989 **

SAMPLE	Acqua nera	IONIC STRENGTH	=	$6.73\ E^{-2}$
CODE	16	SUM OF ANIONS	=	60.39 meq./l
AREA CODE	Apuane	SUM OF CATIONS	=	61.25 meq./l
DATE	01/05/1989	DISCREPANCY	=	1.41 %
TEMPERATURE (^{o}C)	15.0			

	Species	ppm	[M]	log [M]	I	a	log (a)
1	Na^{+}	1196.65	$5.20\ E^{-2}$	-1.2837	0.799	$4.16\ E^{-2}$	-1.3811
2	K^{+}	17.90	$4.58\ E^{-4}$	-3.3394	0.787	$3.6\ E^{-4}$	-3.4434
3	Ca^{++}	115.54	$2.88\ E^{-3}$	-2.5402	0.451	$1.3\ E^{-3}$	-2.8859
4	Mg^{++}	34.93	$1.44\ E^{-3}$	-2.8428	0.489	$7.02\ E^{-4}$	-3.1535
5	HSO_4^{-}		$5.33\ E^{-9}$	-8.2733	0.799	$4.26\ E^{-9}$	-8.3707
6	$NaSO_4^{-}$	7.99	$6.71\ E^{-5}$	-4.1731	0.802	$5.39\ E^{-5}$	-4.2686
7	KSO_4^{-}	0.32	$2.33\ E^{-6}$	-5.6323	0.793	$1.85\ E^{-6}$	-5.7329
8	CO_3^{--}	0.35	$5.82\ E^{-6}$	-5.2353	0.419	$2.44\ E^{-6}$	-5.6128
9	HCO_3^{-}	337.77	$5.54\ E^{-3}$	-2.2568	0.805	$4.46\ E^{-3}$	-2.3511
10	HS^{-}		$4.46\ E^{-3}$		0.793	$4.46\ E^{-3}$	
11	$H_2BO_3^{-}$	0.01	$1.95\ E^{-7}$	-6.7104	0.799	$1.56\ E^{-7}$	-6.8078
12	$H_3SiO_4^{-}$	0.02	$1.61\ E^{-7}$	-6.7937	0.843	$1.36\ E^{-7}$	-6.8677
13	OH^{-}		$8.92\ E^{-8}$	-7.0496	0.793	$7.08\ E^{-8}$	-7.1502
14	NH_4^{+}	0.09	$4.99\ E^{-6}$	-5.3015	0.787	$3.93\ E^{-6}$	-5.4055
15	$CaHCO_3^{+}$	5.6	$5.54\ E^{-5}$	-4.2563	0.812	$4.5\ E^{-5}$	-4.3468
16	$MgHCO_3^{+}$	3.64	$4.28\ E^{-5}$	-4.3685	0.799	$3.42\ E^{-5}$	-4.4659
17	Fe^{++}		$3.42\ E^{-5}$		0.451	$3.42\ E^{-5}$	
18	Cl^{-}	1790.38	$5.05\ E^{-2}$	-1.2967	0.787	$3.97\ E^{-2}$	-1.4007
19	SO_4^{--}	205.32	$2.14\ E^{-3}$	-2.6701	0.408	$8.72\ E^{-4}$	-3.0596
20	H^{+}		$7.65\ E^{-8}$	-7.1161	0.843	$6.46\ E^{-8}$	-7.1900
21	S^{--}		$6.46\ E^{-8}$		0.430	$6.46\ E^{-8}$	
22	Li^{+}	0.14	$2.00\ E^{-5}$	-4.699	0.820	$1.64\ E^{-5}$	-4.7854
23	F^{-}		$1.64\ E^{-5}$		0.793	$1.64\ E^{-5}$	
24	Sr^{++}		$1.64\ E^{-5}$		0.430	$1.64\ E^{-5}$	
25	$CaSO_4$	14.73	$1.08\ E^{-4}$	-3.9658	1.000	$1.08\ E^{-4}$	-3.9658
26	$MgSO_4$	10.23	$8.51\ E^{-5}$	-4.07	1.000	$8.51\ E^{-5}$	-4.0700
27	H_2CO_3	46.64	$8.00\ E^{-4}$	-3.0967	1.000	$8.0\ E^{-4}$	-3.0967
28	H_2S		$8.00\ E^{-4}$		1.000	$8.0\ E^{-4}$	
29	H_4SiO_4	11.13	$1.16\ E^{-4}$	-3.9362	1.000	$1.16\ E^{-4}$	-3.9362
30	NH_4OH		$5.84\ E^{-9}$	-8.2333	1.000	$5.84\ E^{-9}$	-8.2333
31	H_3BO_3	1.41	$2.28\ E^{-5}$	-4.642	1.000	$2.28\ E^{-5}$	-4.6420
32	SiO_2		$2.25\ E^{-8}$	-7.6472	1.000	$2.25\ E^{-8}$	-7.6272
33	$CaCO_3$	0.35	$3.54\ E^{-6}$	-5.4508	1.000	$3.54\ E^{-6}$	-5.4508
34	$MgCO_3$	0.12	$1.45\ E^{-6}$	-5.8373	1.000	$1.45\ E^{-6}$	-5.8373

Example on printer output of activity coefficients and activities calculated in a sample of a secondary extracted file from the main groundwaters data-base.

Table 3. Saturation in calcite, anhydrite, and fluorite

```
** file = ITALIA.WAT ********* S A T U R A T I O N S ******** 25/05/1989 ***

SAMPLE              Abano Terme              log Kps   log Ka
CODE                14                       -------   -------
AREA CODE           Italia          calcite   -9.162    -8.605  (oversaturated)
DATE                01/03/1989    anhydrite   -5.373    -5.244  (oversaturated)
TEMPERATURE (°C)    83.0           fluorite  -10.574   -10.130  (oversaturated)
```

Example on printer output of solubility products for the evaluation of saturation condition in calcite, anhydrite and fluorite, in a sample of a secondary extracted file taken from the main grounwaters data-base.

Table 4. Temperatures (°C) evaluated with chemical geothermometers in liquid phase

** file = ITALIA.WAT *********** G E O T H E R M O M E T R Y ********* 25/05/1989 **

CODE	SAMPLE	T_{calc}	T_{qtz}	$T_{Na-k}1$	$T_{Na-K}2$	$T_{Na-K-Ca}$	$T_{Na-K-Ca-Mg}$	T_{Na-Li}	T_{Li}	T_{K^2-Mg}	T_{Mg-Li}
14	Abano Terme	80	110	182	144	167	82	-	628	96	399
21	Acqua Borra	23	62	178	140	178	88	-	737	117	442
35	Acqua Santa	30	68	158	116	147	69	-	-	81	-
52	Acque Sante	39	73	54	3	80	69	-	-	92	-
54	Ali'	-	-	154	112	161	30	-	-	94	-
1	Aqui Terme	73	104	115	68	98	30	-	479	100	469
36	Bagno Orte	23	62	300	297	33	33	-	344	45	198
24	Bagnolo	71	101	370	402	59	59	738	182	63	108
9	Bormio	46	81	252	233	4	4	503	142	24	59
16	B. Romagna	39	75	97	48	118	78	218	182	79	166
12	Caldiero	31	69	264	249	40	40	-	297	41	187
32	Canino	65	97	353	377	69	69	-	340	73	194
48	Caronte N.	85	113	172	133	158	76	-	454	88	283
50	Caronte S.	65	97	193	158	48	48	894	280	44	147
20	Casciana	48	83	248	227	-4	-4	-	211	19	97

Example on printer output of temperatures computed with chemical geothermometers in liquid phase in samples of a secondary extracted file taken from the main grounwaters data-base.

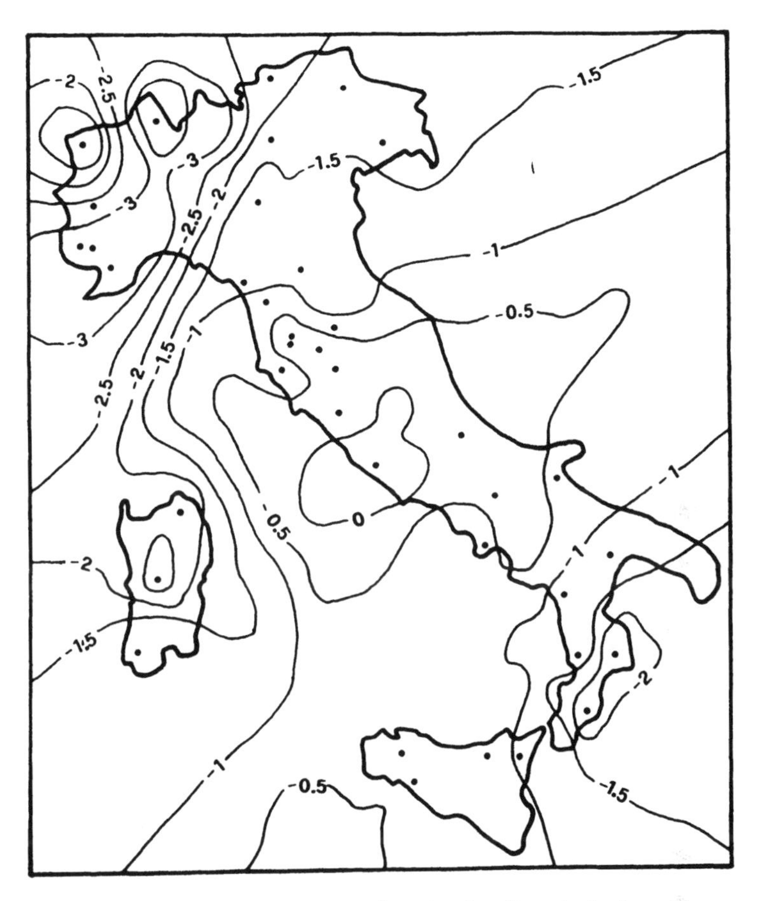

Figure 6. Example of isodistribution map showing log Pco_2 (calculated by program) in Italian thermal springs.

available "Surfer" package of Golden Software—areal or three-dimensional distribution maps of a selected chemical variable (or combinations) in certain areas, as shown in Figure 6. This type of representation of data can be useful particularly in the field of geochemical prospecting, and in monitoring defined areas (e.g., a hydrologic basin, an aqueduct network....etc.), for environment control purposes.

Further details of Hydrodat® can be requested to ECOSYSTEMS s.a.s. Via Mariti 10, 50127 Firenze (Italy).

ACKNOWLEDGMENTS

The authors wish to thank Mr. M. Valenti, Italian National Geothermal Unit (ENEL) of Pisa (Italy) for his valuable help.

REFERENCES

Davis, J.C., 1973, Statistics and data analysis in geology: John Wiley & Sons, New York, 550p.

Garrels, R.M., and Christ, C.L., 1965, Solution mineral and equilibria: Harper and Row, New York, 450p.

Giggenbach, W., 1986, Graphical techniques for the evaluation of water-rock interaction conditions by use of Na,K,Mg and Ca contents of discharge waters: Proc. 8th New Zealand Geothermal Workshop, p. 37-44.

Henly, R.H., Truesdell, A.H., Burton, P.B., and Whitney, J.A., 1984, Fluid-mineral equilibria in hydrothermal systems: Econ. Geology, Review in Economic Geology, v. 1, 267p.

Langelier, W., and Ludwig, H., 1942, Graphical methods for indicating the mineral character of natural waters: Jour. Am. Water Works Assoc., v. 34, p. 335-352.

Merino, E., 1979, Internal consistency of a water analysis and uncertainty of the calculated distribution of aqueous species at 25 °C: Geochim. Cosmochim. Acta, v. 43, no. 9, p. 1533-1542.

Reed, M., and Spycher, N., 1984, Calculation of pH and mineral equilibria in hydrothermal waters with application to geothermometry and studies of boiling and dilution: Geochim. Cosmochim. Acta, v. 48, no. 7, p. 1479-1492.

MINIDENT - SOME RECENT DEVELOPMENTS

Dorian G.W. Smith and Heida Omoumi
University of Alberta, Edmonton, Alberta, Canada

ABSTRACT

MinIdent is a command-driven program developed originally on a mainframe and now ported to a PC. Since its initial description, a substantial amount of new data and several new features have been added. In addition to information published on new mineral species, data have been included for about 700 presently unnamed minerals. A subset facility also has been added, as well as an extensive synonymy which provides abbreviated information and source references for some 1500 synonyms, varieties, discredited minerals and species of dubious authenticity. A full mineral classification scheme also has been included. MinIdent-PC requires at least 560 kbytes of usable RAM, one 32 Mbyte HD drive and a math coprocessor.

INTRODUCTION

The MinIdent database and software for mineral identification have undergone substantial development since their initial description (Smith and Leibovitz, 1984, 1986; Smith, 1986). Importantly, they have been ported successfully to a microcomputer ("MinIdent-PC") and in addition numerous facilities have been added or improved. The database also has been substantially enlarged. These developments, which will be described on the following pages, have been implemented on both the mainframe and PC versions. At the present time both versions are

Use of Microcomputers in Geology, Edited by D.F. Merriam and H. Kürzl, Plenum Press, New York, 1992

command driven, a mode of operation that seems to have retained the preference of a majority of serious scientific users of such software.

GENERAL DESCRIPTION

It has been said that the most important thing to know about a mineral is its name because this allows retrieval of all of the information that has been gathered through the years about that mineral. The MinIdent database is intended primarily for identification of minerals and retrieval of their associated data. A comprehensive list of the properties and parameters included in Minident was given in Smith and Leibovitz (1986). Numerical data consist of compositional, optical, and physical properties. In addition to these, abbreviated literature references, geographic locations, geological occurrence, and relevant remarks - particularly the hand-specimen characteristics - also are included. Summaries of properties are contained in displays. Alternatively, tabulations may be used to show the variations of properties and parameters of interest amongst any number of selected minerals.

DOCUMENTATION

MinIdent is documented thoroughly by comprehensive users' manuals for both the mainframe and the PC versions. These indexed manuals not only offer more complete explanations of the terms and commands that can be used but also provide examples of their use. In addition, extensive sections are included with worked examples illustrating how the MinIdent software may be used for mineral identification and other purposes.

The MinIdent (MTS) manual illustrates how the program and its database can be used with the AMDAHL mainframe computer at the University of Alberta. Access can be arranged to this computer and the MinIdent program using data communications networks such as DATAPAC, TELENET, TYMNET, etc. The MinIdent-PC manual illustrates how the program and its database can be used with IBM-compatible AT clones or 386 machines. For successful operation, such computers must be equipped with at least 560 kbytes usable RAM, one 32 Mbyte (or larger) HD drive, and a math coprocessor.

HELP

In line with most modern computer software, considerable effort has be made to ensure that the MinIdent program is "user friendly." Virtually all commands, procedures, and parameters are explained in an extensive HELP facility which can be accessed at any time from within the program by typing HELP (or ?) followed by the term for which further information is required. A tabulation of all the terms that can be explained in this way can be obtained by simply typing LIST (Table 1). Capital letters in this list indicate the minimum input required.

THE DATABASE

The database consists of a compilation of mineral data from the literature. Four different types of data have been used to construct the database and their current size and their relationship is shown in Figure 1:

SAMPLE DATA: Data specific to one particular sample of a mineral are included in a sample record.

GENERAL DATA: Ranges or generalized values observed for one or more properties of several samples of a given mineral are included in a general record.

MINERAL DATA: Parameters that are constant from sample to sample are included in a mineral record. Examples are: name, formula, symmetry, space group, and classification.

COMPILED DATA: All the sample, general, and mineral records for a given mineral are compiled by the program into a compiled record which is the data set used for identification purposes. The maxima, minima, means, and in the situation of compositions, standard deviations are determined by the program and represent the observed variation of data for each parameter.

USE IN MINERAL IDENTIFICATION

The 4659 minerals shown as being included in the database in Figure 1, include a hundred or so listings for series, classes, groups,

Table 1. Help terms

¶.	$.	*.	-.	?.
:.	#.	@.	A.	ABBreviations.
ACCess rights.	ADD.	ALL.	ALPha.	ANGLE-ENTRY.
ANGLES.	ATTENtion.	AVErage.	B.	BATCH.
BATCH FILE.	BATCHPrint.	BETa.	BIRefringence.	BLAnk.
BREAK.	C.	C(Alpha).	C(Beta).	C(Epsilon).
C(Gamma).	C(Omega).	CALPHA.	CASE.	CBETA.
CEPSILON.	CGAMMA.	CHARGEs.	CLASSify.	CM.
COLLECTion.	COLOR.	COLOR(Alpha).	COLOR(Beta).	COLOR(Epsilon).
COLOR(Gamma).	COLOR(Omega).	COLOUR.	COLOUR(Alpha).	COLOUR(Beta).
COLOUR(Epsilon).	COLOUR(Gamma).	COLOUR(Omega).	COLOUR-ENTRY.	
COLOURS.	COMEGA.	COMMAND.	COMMANDS.	COMMENTS.
COMPILE.	COMPILED Data.	CONTINUATION.	CONTINUE.	COORDination.
COPYRIGHTs.	CRD.	CUTOFF.	D-Values.	DATA.
DATA SOURCES.	DATA Types.	DATE.	DEFAULT.	DEFAULTS.
DELETE.	DENsity.	DESTROY.	DEVELOPMENT CHARGES.	
DICHROISM.	DIMension.	DINitials.	DISCLAIMER.	DISCREDited.
DISPersion.	DISPLAY.	DISPLAYING.	DIVisions.	DOCUMENTATION.
DSOurce.	DVAlues.	EDATA2.	EDGes.	EDIT.
EDITING.	ELIST.	EPSilon.	ERRORs.	EXIT.
EXPLAIN.	FORmula.	GAMma.	GENERAL Data.	GENERALS.
GRECords.	HELP.	ID.	IDENT SECTION.	IDENTIFIER.
IDENTIFY.	IGNOREd.	IM.	INDices.	INITIals.
INTRODUCTION.	JCPds.	LCUTOFF.	LENGTH.	LEVel.
LIST	LOCality.	LOCATion.	MATCH.	MAXimum.
MERGE.	MINERAL.	MINERAL Data.	MINIDENT.	MINIMum.
MISmatch.	MLOcality.	MOCcurrences.	MODIFY Section,MOHs.	
MOVE.	MREMarks.	MTS.	N.	N(Alpha).
N(Beta).	N(Epsilon).	N(Gamma).	N(Omega).	NALpha.
NAMe.	NBEta.	NEPsilon.	NEWMINeral.	NGAmma.
NOMega.	NONe.	NULLs.	NUMber.	OAP.
OCCurrence.	OF.	OMEga.	OPM.	OUTPUT.
OXides.	PASSWORD.	PBASE.	PBASIS.	PERMIT.
PLEOchroism.	POLymorphs.	PRECISION.	PROPORTION-Basis.	
PROPORTION-Sum.	PSUM.	PSW.	Quit.	
R(470).	R(546).	R(589).	R(650).	RANGE.
RECords.	REFERences.	REFLectance.	REFLECTANCE(470nm).	
REFLECTANCE(546nm).	REFLECTANCE(589nm).			
REFLECTANCE(650nm).	REFRaction.	REMarks.	REQuests.	
RESTRICT.	RFL.	RFR.	RIGhts.	R470nm.
R546nm.	R589nm.	R650nm.	SAMPLE Data.	SAMPLEs.
SAVE.	SET.MIN.	SGLIST.	SHOW.	SHOWNull.
SIGN.	SORT.	SOURCE.	SPAce-group.	SPCGRP.
SRECords.	STATUS.	STop.	SUBSET.	SUGGESTions.
SUM.	SYMBOLS.	SYMmetry.	SYNONYM Data.	SYNONYMS.
SYNTAX.	TABLE.	TABULATE.	TEST.	TIMe.
TITLE.	TLIST.	TM.	TREE.	TYPE.
Unknown.	UNKNOWN Data.	UNLOAD.	VARIABLEs.	VERTICAL BAR.
VHN.	Weight.	WIDTH.	YEAr-first-described.	
YFD.	YRECords.	ZERO.	2V.	2V(Alpha).
2V(Gamma).	2V+.	2V-.	2VAlpha.	2VGamma.
2VX.	2VZ.			

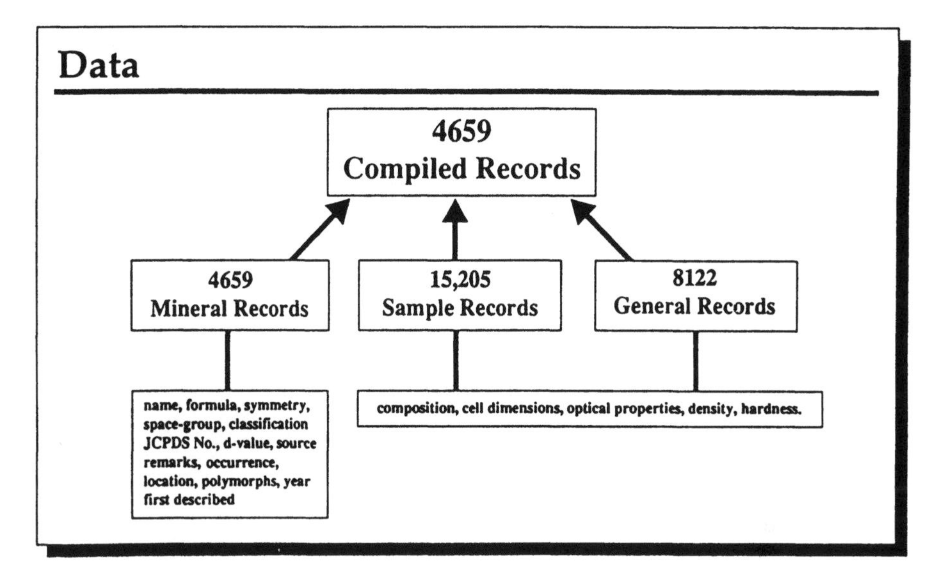

Figure 1. Current size and relationship of data types in database.

families, etc. The remaining ~4,500 minerals (including almost 700 described but presently unnamed minerals) constitute a formidable number of possibilities for the practicing mineralogist attempting to make that all-important first step of mineral identification. There now are many programs in the private, public, and commercial domains which offer assistance in the problems of mineral identification. For the most part these are specific to certain techniques, for example reflectance microscopy - Gerlitz and Leonard (1989); or X-ray powder diffraction (a recent review of programs in this area was given by Smith, 1989). In the vast majority of situations, the software uses simple search/match algorithms. The power of MinIdent for identification purposes comes from its use of data obtained from any of a whole range of techniques and also from its use of "scoring algorithms" (see IDENTIFY commands) to rank minerals according to their similarity to the material being investigated.

The basic command sequence used in mineral identification is as follows:

UNKNOWN
...parameters
SAVE
MATCH
UNKNOWN EDIT

...parameters
SAVE
IDENTIFY AND
TABULATE
DISPLAY

In the list:

UNKNOWN — is used to enter input mode for specifying parameters.

...parameters — represents data that have been obtained for the unknown (e.g., composition, refractive indices, hardness, d-values etc.) and which are to be used in the search.

SAVE — indicates to the program that input of information is complete.

MATCH — searches the entire database for minerals that exactly match the properties of the unidentified mineral. Data for the unidentified mineral usually are entered as ranges which includes generous error limits in order to consider all possible minerals in the match process and allow for possible analytical inaccuracies. The primary purpose of MATCH is to reduce the number of possibilities that need to be considered in detail.

UNKNOWN EDIT — indicates that additional or more precise information for the mineral being identified is to be entered.

IDENTIFY — selects the most likely matches and is different from the MATCH procedure in that it will not eliminate minerals if one or more of their parameters do not exactly match those of the unidentified mineral. Instead, demerit points are assigned and a matching index is calculated for each possible candidate.

...AND — is a "logical and." It may be used in conjunction with either the MATCH or IDENTIFY commands to limit consideration to the list of possibilities generated by the previous procedure.

TABULATE lists, in tabular format, matched or identified minerals together with the corresponding data used in the identification process. It also may be used to tabulate other properties of the matched minerals.

DISPLAY may be used to display the compiled record of any of the matched minerals in order to obtain a complete compilation of all the various parameters for that mineral.

As an example of the application of the identification procedures in a real situation, we shall take a rather poor analysis (low total) of a complex rare earth-bearing uranium mineral and attempt an identification on the basis of the published composition.

```
MinIdent command”””?UNKNOWN
Unknown Input”””?w Nb2O5=major
Unknown Input”””?w TiO2=major
Unknown Input”””?w Y2O3=major
Unknown Input”””?save
MinIdent command”””?MATCH
Scanning within MINERALS division.
4659 minerals examined,   0 ignored, 4637 not matched and   22 matched.
MinIdent command”””?UNKNOWN EDIT
Unknown Input”””?w=null
Unknown Input”””?w U3O8=22.7
Unknown Input”””?w ThO2=6
Unknown Input”””?w Y2O3=18.9
Unknown Input”””?w Nb2O5=26.9
Unknown Input”””?w Ta2O5=0.4
Unknown Input”””?w TiO2=19.7
Unknown Input”””?w FeO=1.9
Unknown Input”””?w Dy2O3=3.9
Unknown Input”””?w Er2O3=2.4
Unknown Input”””?w Ho2O3=1.4
Unknown Input”””?save
MinIdent command”””? IDENTIFY AND
100% of database identified.
22 minerals examined,   0 ignored and the top 20 identified.
MinIdent command”””?TABULATE
Defaults are: unk nam tm w O w Ti w Fe w Y w Nb w Dy w Ho w Er w Ta w Th w U
(See Table 2)
```

In this list, it will be noted that, in the first instance, crude information on composition (major meaning anywhere between 3% and 100%) was entered and the MATCH procedure used. The objective was

Table 2. TM value (total matching index) gives measure of "goodness-of-fit"

Name	TM	O	Ti	Fe	Y	Nb	Dy	Ho	Er	Ta	Th	U
unidentified sample		25.4	11.692	1.462	14.734	18.616	3.364	1.21	2.078	0.324	5.22	19.057
		25.913	11.928	1.492	15.031	18.992	3.432	1.234	2.12	0.331	5.326	19.442
EUXENITE-(Y)	78.3	28.324	8.242	4.379	13.734	27.033	1.18	0.0	0.924	3.78	2.027	7.582
POLYCRASE-(Y)	75.5	26.566	14.285	1.41	13.235	15.465	1.613	0.428	1.582	5.288	3.098	7.749
UM1926-01	52.5	25.383	19.925		12.327	28.118			19.053	35.725		
AESCHYNITE-(Y)	45.8	26.196	16.172	1.811	8.779	17.005	2.63	0.506	1.619	3.57	5.618	2.423
PISEKITE	45.1	25.137	2.518	2.021	7.334	28.661				5.651	1.845	8.815
FERGUSONITE-(Y)	39.4	22.481	1.327	0.777	17.997	25.919	2.86		2.689	3.67	1.13	
AESCHYNITE-(CE)	29.4	25.608	14.834	1.57	1.932	19.592	0.535	0.106	0.308	1.943	11.212	0.632
MURATAITE	29.3	25.378	22.703	3.397	9.497	6.997	1.899	0.498	2.702			
FERSMITE	29.2	29.117	4.003	0.779	1.541	45.338	0.632	0.151	0.511	3.749	0.608	0.476
YTTROBETAFITE-(Y)	27.3	33.371	9.023	2.903	6.189	22.446				6.646	0.923	7.946
UM1930-01	26.5	28.206	2.572	4.092	13.89	28.933				5.954	1.389	10.658
KOBEITE-(Y)	21.9	32.486	20.308	8.382	18.158	3.597					0.883	9.578
UM1971-01	21.0	23.813	13.249	9.017	7.126	5.767					3.603	2.997
SAMARSKITE-(Y)	19.7	26.132	1.572	6.106	5.548	25.651	1.289	0.251	1.461	10.236	2.408	10.883
YTTROPYROCHLORE-(Y)	18.9	32.924	1.788	3.498	10.55	27.651	0.0		0.0	5.219	0.991	7.855
SINICITE	18.9	27.858	7.482	2.819	5.583	23.883				1.126	7.127	5.33
TANTALAESCHYNITE-(Y)	18.7	23.105	7.991	0.684	3.488	12.927	0.534	0.096	0.306	30.717	4.572	0.353
ZIRKELITE	16.0	28.885	17.993	4.958	2.066	4.095	0.469		0.251	0.287	1.932	0.966
NIOBO-AESCHYNITE-(Y)	12.1	23.853	11.451	0.745	3.481	27.197	0.862	0.0	0.331	6.779	9.465	0.339
SCHETELIGITE	11.2	26.88	11.229	1.461	4.725	6.461				16.379		

Note: blanks indicate no data are available. The determined absence of an element is indicated by a zero.

to reduce greatly the number of possibilities that needed to be considered in the much more complex scoring algorithms used by the subsequent IDENTIFY procedure. In Table 2 the TM value (total matching index) gives a measure of the "goodness-of-fit." The values of 78 and 75 for euxenite-(Y) and polycrase-(Y) are not particularly high - indicating to the user that the results should be regarded with some caution. In this particular situation the analyzed sample has an abnormally high U3O8 context - higher than any previously recorded in the literature.

OTHER FACILITIES

A number of special facilities have been developed for MinIdent. These include:

Subsets

Creation of a subset allows a user to consider only a certain restricted group of minerals. Examples of such subsets might include:

all known AMPHIBOLES
all known PYROXENES
all known RARE EARTH-BEARING MINERALS
all known PLATINUM GROUP MINERALS ("PGM's")
all minerals known to occur in METEORITES
all unnamed minerals
any group of minerals of interest to the user

Once created, subsets can be stored and invoked at will. Ultimately they may prove particularly useful for teaching purposes. They also reduce the search time drastically.

Unnamed Minerals

This is a compilation of data for minerals which have occurred in the literature but which have not yet received names. Unnamed minerals are scattered throughout the literature as inadequately described but possibly new species. The MinIdent database contains the first computer compilation of their optical, physical, and chemical properties. Presently there are nearly 700 unnamed minerals in the database, many of which are undoubtedly genuine new species for which a name has not been proposed. The majority are described inadequately minerals whose identification has not been possible based on the data available, and further work is required. Ultimately, the entire suite of unnamed minerals will be matched against the rest of the database in order to determine whether any now may be equated with minerals that are known presently and named.

A scheme has been adopted for the nomenclature of these unnamed minerals consisting of the year first described, followed by an arbitrary number. For example, UM1976-15 represents an unnamed mineral that was described in 1976; it was probably the 15th unnamed mineral entered for that year.

Synonyms and Discredited Minerals

An extensive synonymy presently with nearly 1500 entries has been incorporated into the database. It includes many varietal and archaic names, in addition to spelling variants. In each situation the correct name and source references, have been included and, in many instances, brief comments offered. When a request for information about a mineral which occurs in the synonyms list is made, the program retrieves the data from the records for the correct mineral name. Discredited mineral names are those which subsequent to their publication, have been shown to be unnecessary or inappropriate for various reasons. For example, it may have been shown that a so-called mineral is actually an intimate mixture of two previously described species. The source references and reason for discreditation can be obtained from the MinIdent database.

EXAMPLES

NAME: Daphnite
SYNONYM: Chamosite
REMARKS: An unnecessary name for a variety of chlorite.
REFERENCE: Can. Min. v.13, p.178-180.

NAME: Yttromicrolite
This is a discredited mineral.
REMARKS: Determined to be an amorphous mixture of calcium sulphate, tantalite, and microlite.
REFERENCE: Can. Min. v.25, p.374. Amer. Min. v.67, p.164-165.

Mineral Classification

A primary division into TYPES and SPECIES has been adopted in the construction of the database. The term type is used here to refer to the fundamental division on the basis of chemical compositions, for example, silicates, oxides, sulphides, etc. Further division into class, group, family, supergroup, series, variety, etc. also is possible. The general classification scheme used in the MinIdent database follows closely that adopted by Fleischer (1987, 1989) in his GLOSSARY OF MINERAL SPECIES. There are some differences, which have to do mainly with the level assigned. For example, in MinIdent zeolites are considered a FAMILY whereas in Fleischer (1987) they are shown as a GROUP. Complete implementation of a classification scheme has been impeded by the lack of a universally accepted scheme with defined levels. However, apart from providing the user with information about where a particular name fits in the mineral kingdom, the classification facility in MinIdent should be of value in permitting ranges of properties within the various divisions to be examined.

SOME RECENTLY IMPLEMENTED CLASSIFICATIONS: From time to time the nomenclature of important mineral groups becomes so chaotic that an IMA (International Mineralogical Association) Subcommittee may be set up with the task of making recommendations for simplifications, extinction of names, etc. Thus recently pyroxene nomenclature was studied by a special Subcommittee of the IMA COMMISSION ON NEW MINERALS AND NEW MINERAL NAMES. As a result of its recommendations, the list of accepted pyroxene species names was reduced to 20 (with appropriate adjectival modifiers to indicate compositional varieties - e.g., "chromian jadeite"). As a consequence of that work, more than 100 pyroxene names in the literature were declared "obsolete." MinIdent attempts to implement such official IMA recommendations as soon as possible. In most instances, the obsolete name is entered into the synonym list with a brief explanation and reference. Some recent revisions to the nomenclature used include the following:

AMPHIBOLES - Following the scheme proposed by the IMA Subcommittee on Amphibole Nomenclature (Leake, 1978).

PYROXENES - Following the scheme proposed by the IMA Subcommittee on the Pyroxene Nomenclature (Morimoto and others, 1988).

RARE-EARTH MINERALS - Following the IMA approved scheme in which the dominant REE symbol is appended in parenthesis after the species name, for example, monazite-(Nd) (see, e.g., Bayliss and Levinson, 1988).

THE CHLORITE GROUP - Here the classification is based on the proposals of Bayliss (1975).

Table 3 shows the classification scheme recently recommended for the pyroxene minerals (Morimoto and others, 1988) as implemented in MinIdent. Note that this implementation, allows for the possibility of a particular species being classified under more than one higher level (i.e., series, group, etc.)

FUTURE DEVELOPMENTS

Many developments of MinIdent may be undertaken through the coming years, although the degree that these are reliable will be

Table 3. Classification of pyroxene minerals

```
MinIdent command""?tree pyroxene full

 1 MINERALS top
 2   SILICATES type
 3     INOSILICATES class
 4       PYROXENE-FAMILY family
 5         NATALYITE species
 6         MN-MG-PYROXENES sub-group
 7           KANOITE species
 8           DONPEACORITE species
 9         MG-FE-PYROXENES sub-group
10           PIGEONITE species
11           CLINOENSTATITE-CLINOFERROSILITE-SERIES series
12             CLINOFERROSILITE species
13             CLINOENSTATITE species
14           ENSTATITE-FERROSILITE-SERIES series
15             FERROSILITE species
16             ENSTATITE species
17         CLINOPYROXENE-GROUP group
18           KANOITE species
19           CA-PYROXENES sub-group
20             AUGITE species
21             DIOPSIDE-HEDENBERGITE-SERIES series
22               HEDENBERGITE species
23               DIOPSIDE species
24             DIOPSIDE-JOHANNSENITE-SERIES series
25               JOHANNSENITE species
26               DIOPSIDE species
27             ESSENEITE species
28             PETEDUNNITE species
29           NA-PYROXENES sub-group
30             AEGIRINE species
31             KOSMOCHLOR species
32             JADEITE species
33             JERVISITE species
34           CA-NA-PYROXENES sub-group
35             OMPHACITE species
36             AEGIRINE-AUGITE species
37           SPODUMENE species
38           PIGEONITE species
39           CLINOENSTATITE-CLINOFERROSILITE-SERIES series
40             CLINOFERROSILITE species
41             CLINOENSTATITE species
42         ORTHOPYROXENE-GROUP group
43           ENSTATITE-FERROSILITE-SERIES series
44             FERROSILITE species
45             ENSTATITE species
46           DONPEACORITE species
```

dependent upon the extent to which the database can be made self-supporting. Given a favorable response from possible users, development of the following features is envisaged:

1) Addition of new minerals, significant new data for existing minerals, synonyms, and discredited minerals.

2) Incorporation of new, IMA-approved schemes for mineral nomenclature and classification.

3) Gradual elimination of residual data errors - both those of a literature and a typographical origin.

4) Development of a compact disk version of the database.

5) Implementation of menu-driven operation.

AVAILABILITY

MinIdent-PC now is available commercially through ASTIMEX SCIENTIFIC LTD, 351 Wellesley St. East, Toronto, Ontario, Canada. M4X 1H2

REFERENCES

Bayliss, P., 1975, Nomenclature of the trioctahedral chlorites: Can. Mineral, v.13, pt. 2, p.178-180.

Bayliss, P., and Levinson, A.A., 1988, A system of nomenclature for rare-earth mineral species: revision and extension: Am. Mineral., v.73, no. 3-4, p.422-423.

Fleischer, M., 1987, Glossary of mineral species: The Mineralogical Record, Inc., Tucson, Arizona, 227p.

Fleischer, M., 1989, Additions and corrections to the glossary of mineral species (5th ed.): Min. Record, v. 20, no. 4, p. 289-298.

Gerlitz, C.N., and Leonard, B.F., 1989, Reflectance of ore minerals: search-and-match identification system for IBM PC's using IMA/COM quantitative data file for ore minerals, second issue (abst.): 28th Intern. Geological Congress, Abstracts, v.1, p. 544-545.

Leake, B.E., 1978, Nomenclature of amphiboles: Report of I.M.A. Subcommittee on Amphiboles: Mineral. Mag., v. 42, no. 324, p. 533-563.

Morimoto, N., Fabries, J., Ferguson, A.K., Ginzburg, I.V., Ross, M. Seifert, F.A., Zussman, J., Aoki, K., and Gottardi, G., 1988, Nomenclature of pyroxenes: Am. Mineral., v. 73, no. 9-10, p. 1123-1133.

Smith, D.G.W., 1986, Automation of mineral identification from electron microprobe analyses, *in* Microbeam analysis-1986: San Francisco Press, San Francisco, California, p. 153-156.

Smith, D.G.W., and Leibovitz, D.P., 1984, A computer-based system for identification of minerals on the basis of composition and other properties: 27th Intern. Geol. Congress (Moscow, 1984), Abstracts, v. 5, p. 169.

Smith, D.G.W., and Leibovitz, D.P., 1986, MinIdent: A database for minerals and a computer program for their identification: Can. Mineral, v. 24, pt. 4, p. 695-708.

Smith, D.K., 1989, Computer analysis of diffraction data *in* Bish, D. L., and Post, J. E., eds., Modern powder diffraction: Min. Soc. America, Reviews in Mineralogy, v. 20, p. 183-216.

MICROCOMPUTER APPLICATION OF DIGITAL ELEVATION MODELS AND OTHER GRIDDED DATA SETS FOR GEOLOGISTS

Peter L. Guth
U.S. Naval Academy, Annapolis, Maryland, USA

ABSTRACT

Personal computers have the disk storage capacity, processor speed, and color graphics displays necessary to manipulate and display large gridded data sets recording elevations, bathymetry, gravity, and magnetic information. These gridded data sets can be used for general terrain analysis, specific geologic calculations, or as the base for a geographic information system. Standard microcomputer hardware lets geologists evaluate, compare, and use data sets with hundreds of thousands of values.

INTRODUCTION

Gridded data sets provide a compact, efficient way to store large amounts of data, allowing rapid retrieval of randomly selected data. Microcomputers can access these data and provide high-resolution graphical displays and hardcopy output on dotmatrix or laser printers. Data analysis and display, previously restricted to mainframes or expensive

Use of Microcomputers in Geology, Edited by D.F. Merriam
and H. Kürzl, Plenum Press, New York, 1992

workstations (e.g., Duguay and others, 1989; McGuffie and others, 1989; Verhoef and others, 1990), can be performed on personal computers. Microcomputers can generate color output such as slope maps (Moore and Mark, 1986) or shaded relief maps (Pike and Thelin, 1989). Microcomputers bring the capability to manipulate digital data within the reach of all geologists, with acceptable speed and graphics capabilities.

The MICRODEM program has been evolving since 1986 (Guth, 1986; Guth, Ressler, and Bacastow, 1987; Guth, 1988). Initially designed for terrain analysis using a digital elevation model prepared by the U.S. Defense Mapping Agency (DMA), the program now handles a wide variety of gridded digital data sets. The program requires modest investment in computer hardware, and both executable and source code are available. The discussion that follows focuses on the capabilities of MICRODEM. Other programs can duplicate many of these features, but source code for most other programs with similar capabilities is not available.

MICRODEM runs on microcomputers using the MS-DOS operating system. The program requires 640 kb of memory, and a hard disk for storage of realistic amounts of data, although student problem sets can run on dual floppy drive systems. It uses the standard EGA and VGA monitors (some operations will run on CGA or Hercules screens). Higher resolution output, up to 180 dots per inch, appears on dotmatrix printers (both 8 and 24 pin) and on the Hewlett Packard LaserJet. Third party software can capture screen images and redirect them to a color printer or plotter. A math coprocessor greatly improves many program operations. Input is interactive and menu-driven, using either a mouse or the cursor keys. The program accepts scanned images of maps or aerial photographs from a scanner producing TIFF format files, such as the Hewlett Packard ScanJet.

The VGA monitor has a 16 color mode with 640x480 resolution. The 16 colors can be selected from a palette of about 256,000 colors, which includes 64 shades of red, green, blue, cyan, magenta, yellow, or gray. With sixteen shades selected from the 64 available, excellent results can be achieved in displaying black and white images. Page scanners typically produce 16 color images, which the VGA faithfully depicts, and satellite data such as LANDSAT, SPOT, or AVHRR weather images will convey much of their information when displayed with 16 colors. Although the satellites typically record data in 256 intensity levels (or even 1028 for the AVHRR instrument), most images actually use many fewer intensity levels. The VGA thus provides satisfactory image analysis hardware at low cost, and VGA monitors have become available widely even on machines used only for wordprocessing or spreadsheet manipulation. Some MICRODEM applications, such as shaded relief maps, can use this graphic mode to great advantage.

DIGITAL DATA SET FORMAT

A digital elevation model (DEM) uses an array of numbers to record surface elevation. Because of their wide application in other fields such as geography, hydrography, and cartography, a wide literature and nomenclature has been developed for DEMs (e.g., Yoeli, 1983; McCullagh, 1988). The same principles apply to other geophysical data sets, and the same operations can be applied to them, although users initially may have trouble mentally interpreting a three-dimensional representation of the Earth's magnetic or gravity field. For simplicity I will refer to DEMs, when in fact the discussion applies to all gridded data sets. An alternative to gridded data, triangulated networks, will not be discussed here because the readily available data sets use the gridded format. Actually having data available outweighs any theoretical advantage of another format. McCullagh (1988) discusses relative merits of gridded versus triangulated data when the user must select which to use and then creates the data set.

DEMs usually consist of a regular grid of elevations with x-y coordinates implicit in the data structure. The grid structure saves storage space because only the elevations (field values) need to be stored, and it allows rapid random access. DEMs can use rectangular grids such as Universal Transverse Mercator (UTM), or geographic latitude-longitude grids. Because of convergence of the meridians approaching the poles latitude-longitude grids are not rectangular, although over small areas they may be treated as rectangular. Latitude-longitude grids are easier to use when crossing the boundary of one data set to another, because at the 6° zone boundaries UTM coordinates in adjacent areas cannot easily be reconciled. Some UTM-gridded data sets also clip their data to boundaries from latitude-longitude coordinates, which results in irregular rows and columns; programs using these data sets much convert them to a regular grid and flag the missing values.

By convention most digital elevation models have been organized by columns, with the westernmost column first and the easternmost last. Within rows the values progress from south to north. Thus the first point in the array would be the southwestern corner, and the last point would be the northeastern corner. DMA and the U.S. Geological Survey (USGS) both use this format for their elevation models, and the U.S. National Ocean Survey (NOS) has adopted it for its bathymetric data sets. In the terminology of Hittelman, Kinsfather, and Meyers (1989) this is BTR (bottom to top then right), and MICRODEM requires data sets to have this format. Some operations, such as gridded contouring or slope map calculation, require access to several columns of data at a time for efficient computation. Although these operations could work equally well with rows, such flexibility would require different code for the two situations. Conversion of the data to a single format allows simple, fast, single case processing, and MICRODEM will convert from the LRD (left to right then down) format used by the binary output from the Geophysics of North America CD-ROM (Hittelman, Kinsfather, and Meyers, 1989).

Integers usually represent the data values in DEMs. An integer requires only two bytes of computer storage, and can represent values from 32767 to -32768. In contrast storage of values as real number requires four bytes or more depending on the accuracy desired, at least doubling the storage requirements. For elevations and bathymetry using integer values in meters provides sufficient accuracy. Geophysical values with smaller ranges can be stored as integers with appropriate units; magnetic anomalies use tenths of nannoteslas, whereas gravity anomalies use tenths of milligals.

AVAILABLE DATA

Gridded data sets tested extensively with MICRODEM include a variety of types of data. Digital elevation models include the DMA 1:250,000 DEM (available through USGS) and the USGS 1:24,000 DEM (Fig. 1). The DMA data has 3 arc second latitude-longitude spacing, which translates into 60-90 m (variable with latitude) between data points and provides complete coverage of the United States. The basic unit for this DEM, a 1°x1° region, contains 1.44 million elevations. The USGS data has 30 m UTM spacing, but covers only about 25% of the United States. Its basic unit, a 7 1/2'x7 1/2' quadrangle, contains about 175,000 elevations (variable with latitude). NOS has begun producing a digital bathymetric data set covering the Economic Exclusive Zone (EEZ), with data spacing of 250 m on the UTM grid (Fig. 2). The ETOPO5 data set, with both land topography and marine bathymetry at 5 arc minute spacing, covers the entire world.

Digital geophysical data sets on compact disk (Hittelman, Kinsfather, and Meyers, 1989) include 5' topography and bathymetry (from the ETOPO5 data set), 30" topography (thinned from DMA data) and bathymetry, 2.5' gravity and magnetics, and AVHRR satellite images.

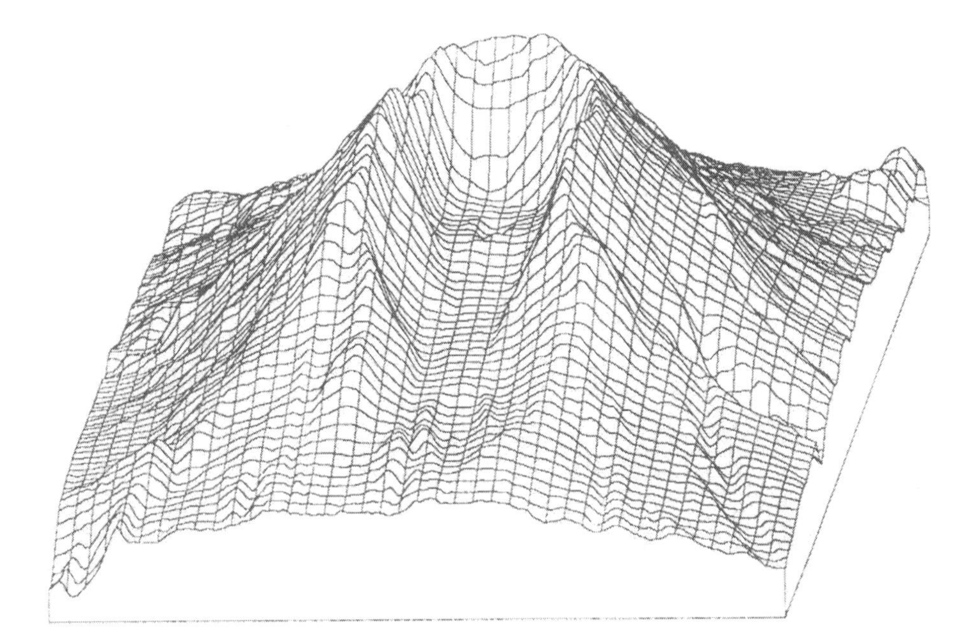

Figure 1. Oblique view of Mt St Helens, Washington, looking south. Data from USGS 1:24,000 DEM. Produced on Hewlett Packard LaserJet with high-resolution software driver.

Some of the data sets cover only the United States, whereas many cover the entire North American plate. Many of the data sets on the CD-ROM represent recent compilations for the Decade of North American Geology.

The computer stores data in binary format as a file of two byte integers. MICRODEM handles variably sized grids; a header file contains the grid size, extreme values, and information on the ellipsoid and datum used for digitization and display. Missing data flags alert the software to avoid spurious interpolations or program crashes. The data sets are limited currently to 1201 rows and an unlimited number of columns, but this could be changed by redefining a single constant and recompiling the program. Larger data sets exceed the display resolution of microcomputer screens, and will produce best results when subdivided. Data sets with fewer than 120,000 values run much faster because the entire data set can load into computer RAM, significantly faster than even a RAM disk.

USGS 1:24,000 DEMs typically have around 175,000 points and require about 350 kb of disk space. A 15'x15' subset of DMA Digital

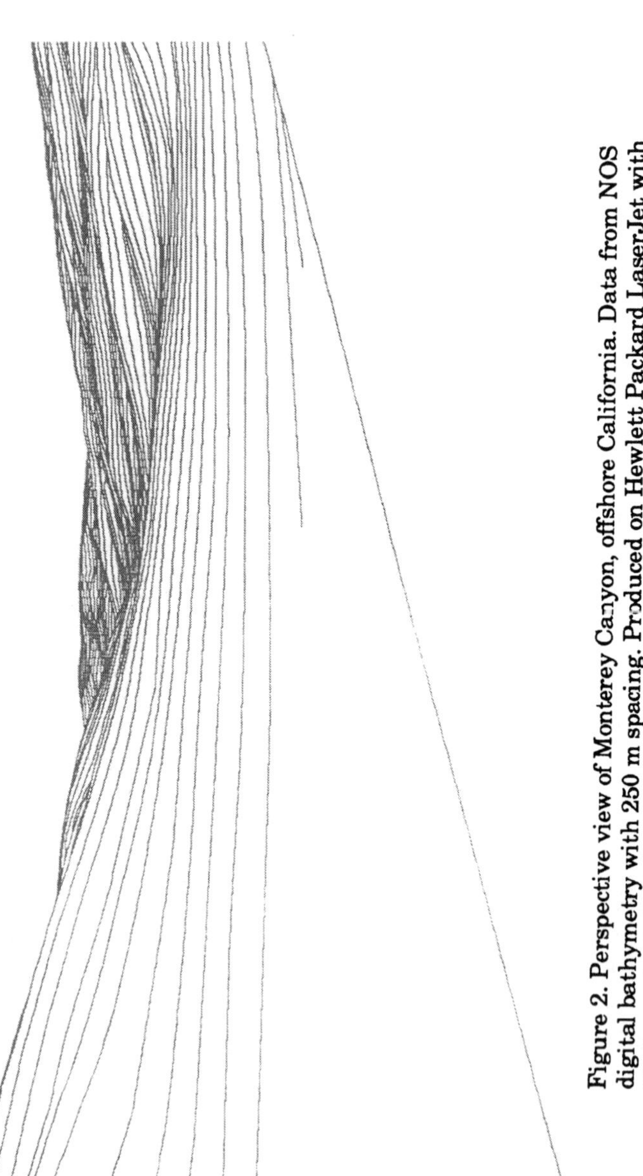

Figure 2. Perspective view of Monterey Canyon, offshore California. Data from NOS digital bathymetry with 250 m spacing. Produced on Hewlett Packard LaserJet with High-resolution software driver.

Terrain Elevation Data (DTED Level 1, the USGS 1:250,000 DEM) has 90,000 points and requires 180 kb while a 30'x30' subset needs 720 kb for 360,000 points. The program considers these three data sets as standards, and file names define the area covered and hence the adjoining data sets. The program can open up to nine adjacent data files and access them to produce output that crosses boundaries between data sets. Public domain or other compression programs typically compress data sets by a factor of about 50% when not in use, saving on disk space.

ACCURACY AND RESOLUTION

Users of a DEM must consider data resolution (spacing) and accuracy (relation of values to "true" elevation). Increasing resolution increases production costs for better photographic coverage of the input stereographic model and more processing. Better resolution also increases storage needs and the number of calculations when using the DEM, important considerations on a microcomputer. Halving the data spacing quadruples the volume of data.

Many users will not have digitization capabilities, especially for the effort involved to create a DEM with tens or hundreds of thousands of data points, and must assess available DEMs to determine suitability for proposed needs. Users designing DEM specifications must recognize the tradeoff between resolution and accuracy (cost) versus area covered, and determine whether full coverage at a small scale is better than partial coverage at a larger scale. Should a DEM not provide absolute quantitative results because of limitations in accuracy or resolution, the computer can provide yet an acceptable qualitative sense of the terrain. As discussed next, MICRODEM contains tools for users to assess the adequacy of a DEM.

For a user without recourse to digitizing (and these large DEMs represent major investments in digitization), a DEM either exists covering the area of interest or does not. If the DEM exists, it is either accurate enough to use or not. That decision will be subjective; in some situations, "pushing" a DEM beyond its accuracy may be preferable to the alternative of using a paper map and doing the work by hand. The user must take responsibility for specific uses of the data. Perfect DEMs will never exist, and there will be tradeoffs in their design and use.

DATUMS

Geologists using DEMs must be aware that coordinates, whether UTM or latitude-longitude, depend upon the datum and ellipsoid used for the projection (Snyder, 1987). Different datums are used throughout the world, both because of irregularities in the Earth's shape that result in locally optimal results assuming different shapes for the Earth, and historical vagaries of mapping from different local starting points. Satellite geodesy has allowed the creation of datums that can be applied worldwide. Changing the datums can result in the same coordinates representing points on the ground separated by several hundred meters.

Datum shifts will soon be a concern with all mapping in the United States. The country is shifting from the North American Datum of 1927 (NAD27) to the North American Datum of 1983 (NAD83) with shifts up to 100 m (Dewhurst, 1990) in locating points on the ground. All coordinates should be prefaced with the datum used to avoid ambiguity and allow precise relocation of points, as paper topographic maps with NAD83 will begin replacing the older maps on NAD27. Even before the datum shift, users of DEMs had to worry about the datums as the different scale DEMs used for a different datum. The USGS 1:24,000 DEMs were digitized on NAD27, whereas the 1:250,000 DEM produced by DMA used the World Geodetic Survey 1972 (WGS72) datum. Newer DEMs in those series will use NAD83 and WGS84 respectively; fortunately the datum of the two are essentially identical. Users of DEMs wanting precise locations must know the datum of their data; with datum shifts up to 100 m, the 1:24,000 DEM can be off up to three or four data points if the wrong datum is assumed, and the 1:250,000 DEM can be off by one data point. Small-scale data sets, similar to 5 arc minute topography, assume a spherical Earth and datum shifts of 100 m are negligible when data spacing reaches 6-9 km.

By default, MICRODEM assumes constant data spacing, using the average along the edges of the data set, on a UTM grid and computes coordinates in a simple rectangular reference frame. For data sets digitized on the UTM grid, such as the USGS 1:24,000 DEM and the NOS EEZ bathymetry, this is exact. For other data sets, similar to the DMA 1:250,000 DEM, this leads to misregistration of about 1% within a 15'x15' region because of the assumption of constant data spacing. Given the other limitations on the 1:250,000 DEM, this additional distortion should be acceptable. For the small-scale data sets such as those with 5 arc minute spacing, the distortion from forcing the data on a regular rectangular grid can be severe, especially at high latitudes. In those situations exact projection on a number of different projections (Sinu-

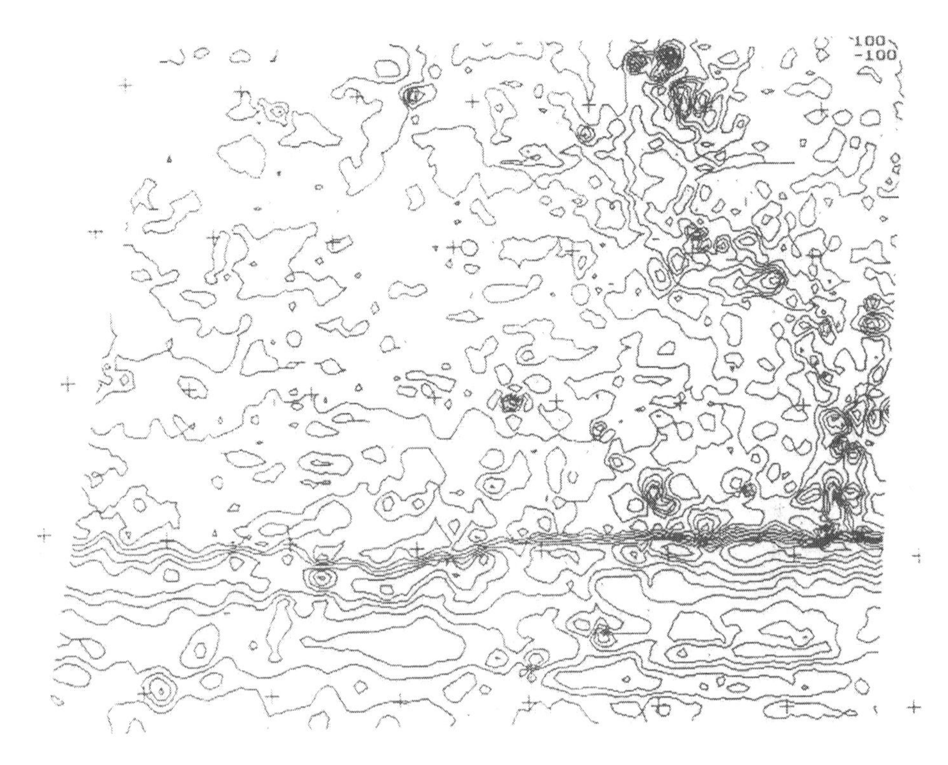

Figure 3. Contour map of gravity field in Pacific. Major discontinuity in southern part of map is Mendocino Fracture Zone. Plotted on Lambert Azimuthal Equal Area projection, apparent from orientation of crosses every 2° of latitude and longitude. Data from NOAA (Hittelman, Kinsfather, and Meyers, 1989). Produced as screen dump of VGA screen on LaserJet.

soidal, Hammer, Orthographic, Lambert Azimuthal Equal Area, Mollweide, Gnomonic, Stereographic, and Mercator) can be used, at a significant penalty in performance to calculate all coordinates (Fig. 3). Currently the precise projections assume a spherical earth, but could be modified for any desired ellipsoid (Snyder, 1987).

IMPORTING DATA

Getting data onto the microcomputer presents one of the greatest challenges to increasing usage of gridded data. Many agencies that create and archive digital data gear their efforts toward nine track magnetic tape, a medium that is not especially available on personal computers. Because of the megabyte and greater size of the data sets, the transfer of large quantities of data becomes a major challenge.

Nine track tape units can be attached to microcomputers (at a cost perhaps equal to or several times that of the computer itself), and provide extremely rapid downloading of data from tape with rates of several megabytes per minute. A cheaper, but slower method involves reading the tape on a mainframe or minicomputer, and then downloading the data onto the microcomputer using a communications program. Transfer rates will depend on the speed of the communications lines available (4800 or 9600 baud represent minimums acceptable), data-line quality, and the degree of error checking. At best I have been able to download several megabytes per hour, and usually several megabytes requires an overnight transfer which runs unattended on an otherwise unused personal computer.

Data are becoming available on other formats usually for the microcomputer. NOS provides their gridded bathymetric data sets on high-density floppy disks, and a commercial firm advertises USGS DEMs on floppy disks (Miller, 1990). Read only compact disk storage units provide cost effective storage for hundreds of megabytes of data, and the National Oceanic and Atmospheric Administration (NOAA) has produced a disk covering the geophysics of the North American tectonic plate (Hittelman, Kinsfather, and Meyers, 1990). Although access to the CD-ROM is significantly slower than a hard disk, the ease of use easily eclipses nine track tape, the only alternative for obtaining such large data sets.

MICRODEM contains import software for three types of data sets, digital elevation models supplied by USGS, NOS digital bathymetric data, and data sets exported from the Geophysics of North America CD-ROM. The USGS DEMs come only on nine track tape in similar ASCII formats; the 1:250,000 DEM covers 1x1 cells in approximately 10 MB files, whereas the 1:24,000 DEM covers 7'x7' quadrangles that require about 1 MB. Once the data have been transferred to the personal computer, MICRODEM will convert the ASCII file into a binary, random access data file, plus create the required header file with minimal input from the user. The NOS bathymetric data sets require significant user input in creating the header, because the header information on the distribution disks does not have a standard format comparable to the USGS DEMs.

The NOAA CD-ROM comes with software to export data into files on the user's hard disk. Because the data sets on the CD- ROM come in a variety of formats, and because of the slow speed of access to the CD-ROM, MICRODEM does not attempt to access directly the CD-ROM. MICRODEM will convert binary files exported from the CD-ROM and automatically create the header; the files are converted to the BTR format used by MICRODEM instead of the LRD format used by the data exported from the CD-ROM (Hittelman, Kinsfather, and Meyers, 1989).

GEOLOGIC APPLICATIONS

DEM applications include general terrain analysis, various geologic operations, and geographic information systems (GIS). General terrain analysis functions available in MICRODEM include colored tinted maps, contour maps, three-dimensional views, and histograms of the elevation distribution. The program also can perform some coordinate transformations, between latitude-longitude and UTM, and between the datum of different data sets. These and related operations on the DEM have application for geomorphology (e.g. Evans, 1980; Franklin, 1987; McCullagh, 1988; Pike, 1988) and terrain evaluation preliminary to any geologic work.

Properly scaled colored tinted maps can display elevation, slope, aspect (the downhill direction), or reflectance. Elevation maps display most rapidly because no calculations need to be performed on the integer data values; the other operations require floating point arithmetic. For simplicity and speed the slope calculation uses the largest of the slopes from the point to its eight nearest neighbors; alternatively a surface could be fitted in the neighborhood of the point and the derivative taken. The aspect map merely calculates slopes in the eight principal cardinal directions and displays a color showing the direction with the maximum downhill slope. The reflectance map uses the algorithm of Pelton (1987) for rapid display of realistic images on a VGA monitor.

Three-dimensional views include both oblique (block diagram, see Fig. 1) and perspective (view from a point, see Fig. 2). Users can select from a number of oblique methods: a painters algorithm which starts at the rear and shows hidden terrain before drawing over it, and hidden-line algorithms that start at the front and cannot show hidden terrain. The hidden-line algorithm can use fishnet tie lines, and optionally fill the blocks in the fishnet with colors tied to elevation. Finally, a rotating oblique revolves in real time on the display screen, allowing the user to see the selected terrain from all sides.

MICRODEM allows comparison of two elevation models covering the same area, or multiple data sets such as bathymetry, gravity, and magnetics. The comparison can take a number of forms, such as superposed profiles (Fig. 4), elevation histograms (Fig. 5), contour maps, or maps showing the difference between values interpolated in two data sets. The comparison allows determination of the accuracy of DEMs

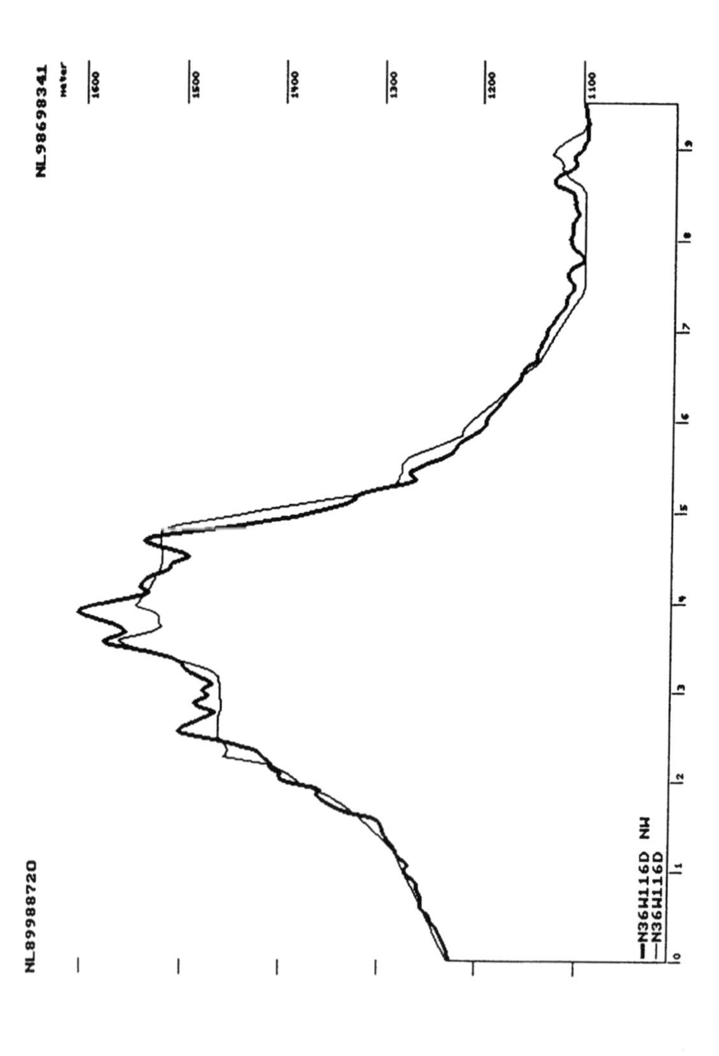

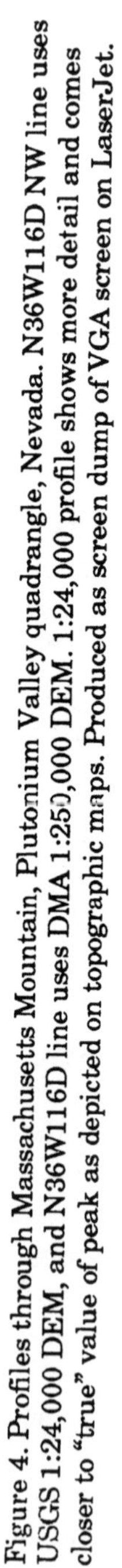

Figure 4. Profiles through Massachusetts Mountain, Plutonium Valley quadrangle, Nevada. N36W116D NW line uses USGS 1:24,000 DEM, and N36W116D line uses DMA 1:250,000 DEM. 1:24,000 profile shows more detail and comes closer to "true" value of peak as depicted on topographic maps. Produced as screen dump of VGA screen on LaserJet.

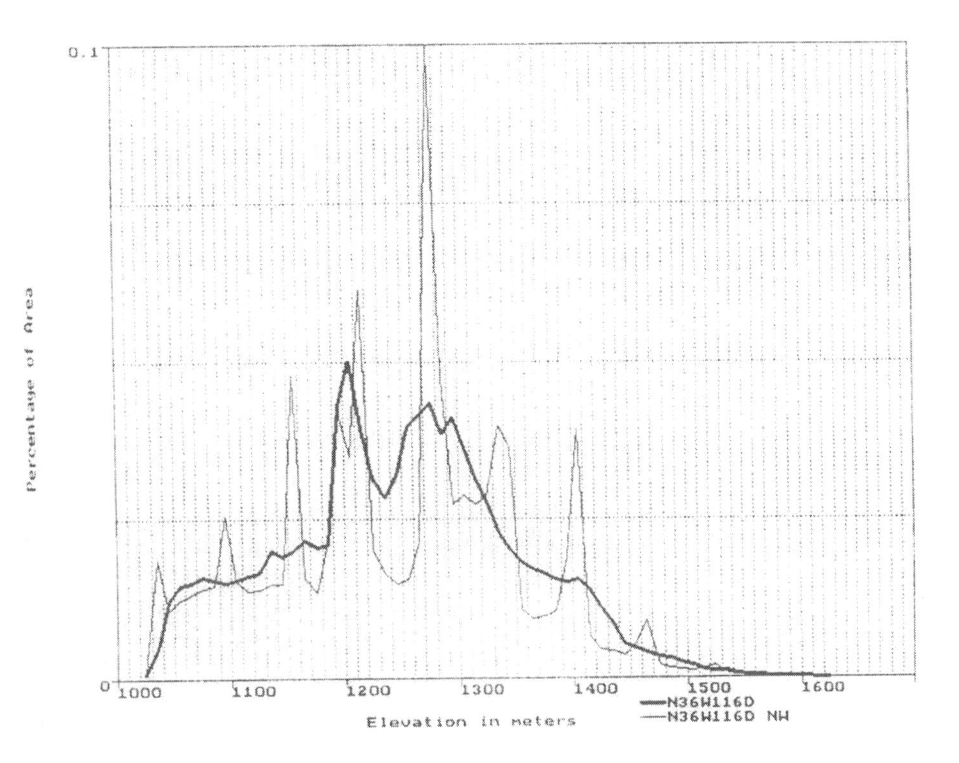

Figure 5. Elevation histograms of Plutonium Valley quadrangle, southern Nevada, from USGS 1:24,000 DEM (N36W116K NW) and corresponding subset of DMA 1:250,000 DEM. Smaller scale 1:250,000 DEM contains spikes corresponding to source map's contour lines, an artifact of digitization. Produced as screen dump of VGA screen on LaserJet.

covering the same area at different scales. Overlaying contours on a scanned image of the map sheet, imported from the HP ScanJet, clearly shows the user the effects of resolution and accuracy of the DEM compared to the paper map.

Hydrologists have used DEMs extensively for basin delineation and storm runoff calculation. MICRODEM incorporates algorithms of Marks, Dozier, and Frew (1984), Jensen (1985), and Martz and DeJong (1988), but probably will require programming modification for optimal results to adjust for the scale of the DEM used and the morphology of the basins investigated.

Guth (1988) discussed using the DEM to help draft geologic cross sections; the CROSSX program now has been incorporated into MICRODEM. DEM use for cross-section generation lets the computer draw topographic profiles and assist drawing by keeping track of many details. Specific geologic applications include using the DEM to solve three-point problems, calculate thicknesses based on map outcrop patterns, or trace the path of a plane (fault or formation contact) across topography. These applications involve repetitive, simple mathematical operations, but in the situation of contact tracing the volume of calculations realistically precludes manual operation. Because of the ease of computer operation, many calculations can be performed rapidly and results compared.

The three-point problem uses the x, y, and z coordinates of three points to determine the orientation of the plane that contains them. After the user selects the three points and the computer interpolates elevations from the DEM, the mathematics is straightforward: calculate the equation of the plane, derive the normal, and convert the normal to standard geologic notation (N35E 75NW) as well as the slope in percent.

The thickness of a unit can be calculated given the strike and dip of a unit and the location of two contacts (upper and lower) oriented along the dip direction. The user selects a point and enters the unit's orientation in one of the following formats: N45E 23SE (strike, Dip, Dip Direction), 23 S45E, 23 135, 23/135, or 23-135 (all variations with dip and dip direction). The computer calculates and draws the dip direction, and the user moves a cursor along the line dip direction to select the upper and lower contacts of thc unit. The computer then calculates the thickness.

If a unit contact or fault is assumed planar, the computer can calculate and graph its outcrop trace on the Earth's surface (Fig. 6). The user interactively selects one point on the contact and enters the orientation of the contact. The search can be a fast threading or an exhaustive search if other, disconnected outcrops of the plane are suspected.

Starting with the known point on the contact, the algorithm considers the four surrounding data points which form a rectangle on the map and a prism in space. Each side of the rectangle corresponds to a line in space, and the computer calculates the general equations of that line (equations of two distinct planes containing the line). The computer then calculates the intersection of the line and the contact from the intersection of three planes, and determines if the intersection lies along the perimeter of the rectangle. Two of the four intersections will lie on the rectangle's perimeter, indicating the surface trace of the contact. The computer connects those points, and follows the contact into the adjacent rectangle. Threading of the contact continues until the contact leaves the screen or closes on the starting point.

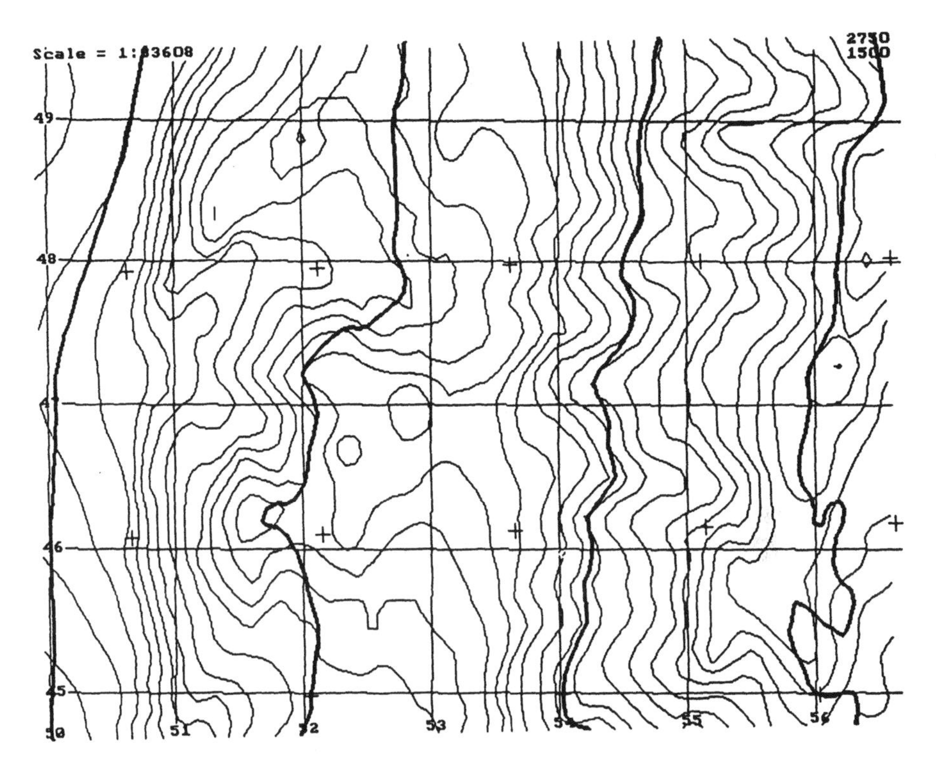

Figure 6. Projected traces of dipping contacts on topographic base map. Contour interval is 50 m. Rectangular UTM grid with 1 km squares is shifted slightly east from true north. Three heavy lines have same orientation (N15E 15E), and were projected from points along southernmost east-west UTM grid line. Produced as screen dump of VGA screen on LaserJet. This figure and next three use part of DMA 1:250,000 DEM for Sheep Range in southern Nevada.

This threading algorithm will not determine other occurrences of the same contact within the map area; at a cost in execution time, they can be located by checking all grid rectangles on the map. The exhaustive search will square approximately the time required for the simple threading of the known contact.

A DEM can serve as the basis for a GIS, particularly appropriate for many geologic applications where surface topography plays a role. A GIS combined with DEM can store well and drill-log data. The program will track well locations, and a number of formation tops specified by the user. The user enters data interactively, the program stores it in an ASCII file, and can display it on a contour map of the topography. It also will project well data onto a line of section (Fig. 7) and show formation tops in cross-section view. Isopachous and structural contour maps can be displayed, overlaid on the topographic map or as a separate display (Fig. 8). A triangulation contouring algorithm (Watson, 1983) works with irregularly spaced data such as well locations. In addition to well data, for which it was designed initially, other data such as sea-surface temperature and barometric pressure for a storm can be plotted and contoured on the topographic or bathymetric base map.

PROGRAM MECHANICS

The program is in Turbo Pascal, version 5.5, and currently includes over 36,000 lines of code. Of this about 7000 lines are the user interface (graphics, printer control, menus, mouse operations) which is available only as a compiled unit. The remainder of the source code, except for experimental modules under development, is available. The executable program is about 200 kb in size, with a 350 kb overlay file that remains

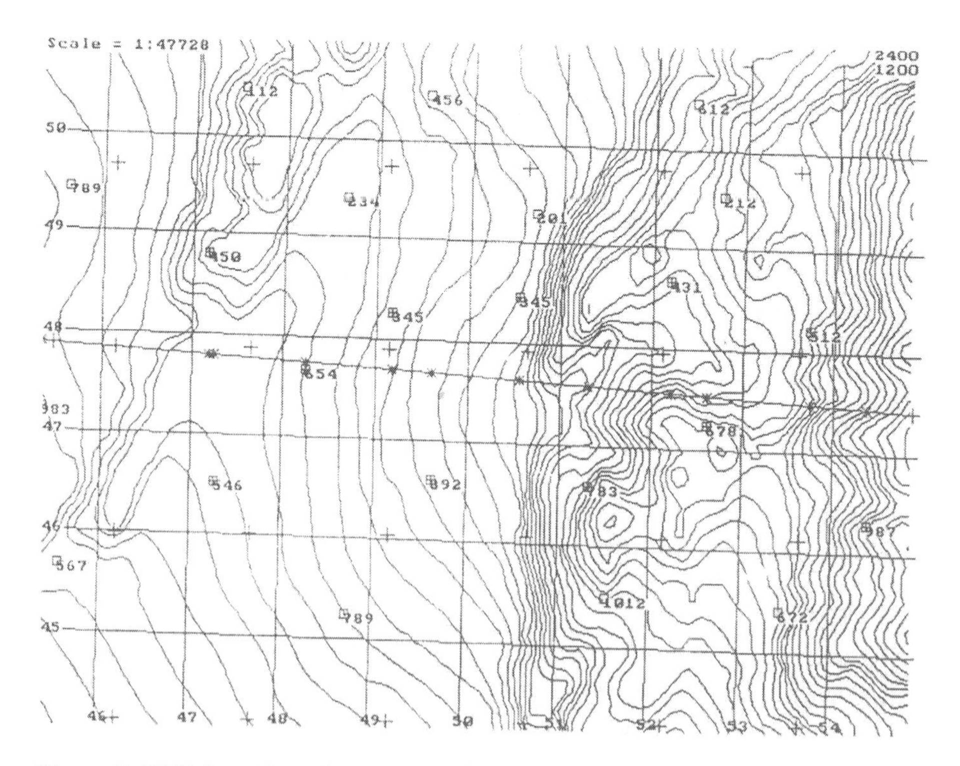

Figure 7. Well data plotted on topographic base and projected onto line of section. Data points indicated with boxes, and those with cross and box are projected onto section line (stars); user specified maximum distance from line to project.

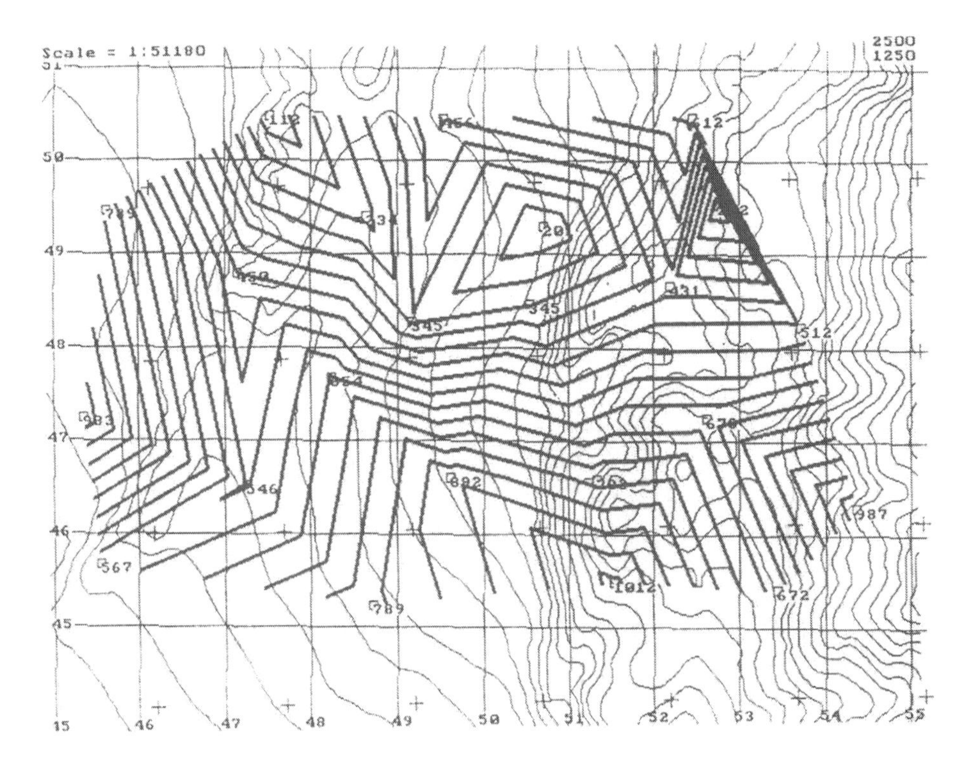

Figure 8. Triangulation contours overlaid on contour base map. Heavy lines show contours of user's data field.

on disk during program operation. The program brings the overlaid major modules into memory only when required, to minimize memory requirements. The program takes up about 200 kb, the digital data set can require up to 240 kb if loaded into RAM for best operations (larger data sets must remain on disk), the printer image consumes 60 kb, and some additional memory is needed for temporary buffers. With current version of MS-DOS easily taking up to 100 kb when loaded with all the necessary device drivers, the 640 kb available to DOS fill rapidly.

FUTURE DEVELOPMENTS

Work in progress focuses on a number of enhancements. Creation of data sets with a scanner or digitizer will make it possible for users in other parts of the world, without available data, to create their own data sets. Field geologists could scan their field map, and use the computer and the DEM to check geometry of bedded units by assuming a dip and calculating thickness at multiple points along an outcrop band, or by calculating dip from three exposures. While DEM accuracy and resolution must be addressed, the geologist could perform a great number of calculations extremely rapidly and should be able to assess accurately at least the magnitude of the values. MICRODEM will superimpose DEM elevation contours on satellite images or scanned aerial photographs displayed on a VGA monitor, or drape satellite images on a three-dimensional block diagram. The program also will be able to work with multiple data sets covering the save area, such as marine bathymetry, gravity, and magnetics, showing all three on screen at the same time.

ACKNOWLEDGMENTS

The initial version of the program was written at the Computer Graphics Laboratory, U.S. Military Academy, to support the Army's terrain analysis teams; it has been expanded greatly since then, partially supported by the Volunteer Program, Office of Regional Geology, U.S. Geological Survey; the Department of Oceanography, U.S. Naval Academy; and the Department of Geography and Environmental Engineering, U.S. Military Academy. Any use of trade names within this paper is for descriptive purposes only and implies no official endorsement.

The program, including source code, documentation, a variety of sample data sets, and an executable version requiring no knowledge of Pascal, is available on seven double-sided, double-density disks for $25 to cover the cost of disks, duplication, and air mail postage.

REFERENCES

Dewhurst, W.T., 1990, The impact of the North American Datum on 1983 on cartographic products: Federal Digital Cartography Newsletter, no. 11, p. 6.

Duguay, C., Holder, G., LeDrew, W., Howarth, P., and Dudycha, D., 1989, A software package for integrating digital elevation models into the digital analysis of remote-sensing data: Computers & Geosciences, v. 15, no. 5, p. 669-678.

Evans, I.S., 1980, An integrated system of terrain analysis and slope mapping: Zeit. für Geomorph, N.F. Supple. Bd. 36, p. 274-295.

Franklin, S.E., 1987, Geomorphometric processing of digital elevation models: Computers & Geosciences, v. 13, no. 6, p. 603-609.

Guth, P.L., 1986, TERRANAL: Microcomputer terrain mapping package, ACSM-ASPRS (American Congress on Surveying and Mapping —American Society of Photogrammetry and Remote Sensing) Annual Meeting, March 1986, Washington, conference proceedings, v. 1, p. 114-122.

Guth, P.L., 1988, Microcomputer-assisted-drawing of geologic cross sections: Jour. Math. Geology, v. 20, no. 8, p. 991-1000.

Guth, P.L., Ressler, E.K., and Bacastow, T.S., 1987, Microcomputer program for manipulating large digital terrain models: Computers & Geosciences, v. 13, no. 3, p. 209-213.

Hittelman, A.M., Kinsfather, J.O., and Meyers, H., 1989, Geophysics of North America CD-ROM Users manual release 1.0: U.S. Department of Commerce.

Jensen, S.K., 1985, Automated derivation of hydrologic basin characteristics from digital elevation model data, *in* Digital representation of spatial knowledge: Auto-Carto 7 Proceedings, p. 301-310.

Marks, D., Dozier, J., and Frew, J., 1984, Automated basin delineation from digital elevation data: Geo-Processing, v. 2, no. 3, p. 299-312.

Martz, L.W., and DeJong, E., 1988, CATCH : a FORTRAN program for measuring catchment area from digital elevation models: Computers & Geosciences, v. 14, no. 5, p. 627-640.

McCullagh, M.J., 1988, Terrain and surface modelling systems: theory and practice: Photogrammetric Record, v. 12, no. 72, p. 747-779.

McGuffie, B.A., Johnson, L.F., Alley, R.E., and Lang, H.R., 1989, IGIS Computer-aided photogeologic mapping with image processing, graphics and CAD/CAM capabilities: Geobyte, v. 4, no. 5, p. 8-14.

Miller, B. (compiler), 1990, Geosoftware: Digital elevation models: Geotimes, v. 35, no. 8, p. 28.

Moore, J.G., and Mark, R.K., 1986, World slope map: EOS [Transactions, American Geophysical Union], v. 67, no. 48, p. 1353, 1360-1362.

Pelton, C., 1987, A computer program for hill-shading digital topographic data sets: Computers & Geosciences, v. 13, no. 5, p. 545-548.

Pike, R.J., 1988, The geometric signature: quantifying landslide-terrain types from digital elevation models: Jour. Math. Geology, v. 20, no. 5, p. 491-512.

Pike, R.J., and Thelin, G.P., 1989, Shaded relief map of U.S. topography from digital elevations: EOS [Transactions, American Geophysical Union], v. 70. no. 38, p. 843,853.

Snyder, J.P., 1987, Map projections—a working manual: U.S. Geol. Survey Prof. Paper 1395, 383 p.

Verhoef, J., Usow, K.H., and Roest, W.R., 1990, A new method for plate reconstructions: the use of gridded data: Computers & Geosciences, v. 16, no. 1, p. 51-74.

Watson, D.F., 1982, Accord: Automatic contouring of raw data: Computers & Geosciences, v. 8, no. 1, p. 97-101.

Yoeli, P., 1983, Digital terrain models and their cartographic and cartometric utilisation: The Cartographic Journal, v. 20, no. 1, p. 17-22.

REUSABLE CODE WORKS!

Fred J. Gunther
Computer Sciences Corporation, Laurel, Maryland, USA

ABSTRACT

Personal experience in microcomputer applications programming for statistical analyses and graphic displays has proven the value of reusing code. Reusing code has been determined to increase software reliability, increase programmer productivity, and decrease development cost.

INTRODUCTION

During the past several years, the author has used a series of data analysis computer programs in his personal work and professional research. For each new program, rather than develop a complete set of new requirements, a new design, a new data structure, a new file structure for disk storage of data, completely new code, a new test plan, and new documentation, the author has reused materials and concepts that he developed in earlier programs. Rewriting and modifying documentation and code modules to provide new or different functions and capabilities from a base of previously developed, well-used and operationally tested modules has resulted in new, reliable code produced quickly, at low cost. The author has determined it useful to reuse code.

Use of Microcomputers in Geology, Edited by D.F. Merriam
and H. Kürzl, Plenum Press, New York, 1992

The programs have performed a variety of functions:

- Data Editor
- Basic Sample Statistical Analysis
- Time-Series Graphic and Statistical Analysis
- Paired-Sample Statistical Analysis (Gunther, 1982)
- Scatter-Diagram (X,Y) Graphic and Statistical Analysis (Fig. 1)
- Bar-Graph Graphic and Statistical Analysis
- Calibration-Pulse Analysis for Landsat TM Data (Gunther, 1984)
- Rose-Diagram Graphic and Statistical Analysis (Gunther, 1986) (Fig. 2)
- Plot Geographic Data on Map Background (Gunther, 1987)
- Time-Duration Graphic and Statistical Analysis
- Age-Pyramid Graphic and Statistical Analysis (Fig. 3)
- Oceanography Hydro-cast Analysis (Fig. 4)
- Oceanography Data Analysis

FEATURES ASSISTING REUSE

The programs are written in a language for a standard, available hardware configuration. They are written in Applesoft BASIC to run on the Apple-II series of microcomputers; two programs have been translated to an Amiga microcomputer. The programs were developed on an Apple II+, with 48K RAM on the main board and 16K RAM on an Apple Language System card in slot 1; programs have been run on Apple IIe and IIc computers. Results are displayed on color or monochrome monitors using normal Apple-II high-resolution graphics. Graphics are printed using the OrangeMicro Grappler interface card to dump the high resolution memory pages to an Epson dotmatrix printer. Statistics (Fig. 5) and annotated data matrices (Fig. 6) are printed on the Epson using standard Apple II system and Applesoft BASIC commands.

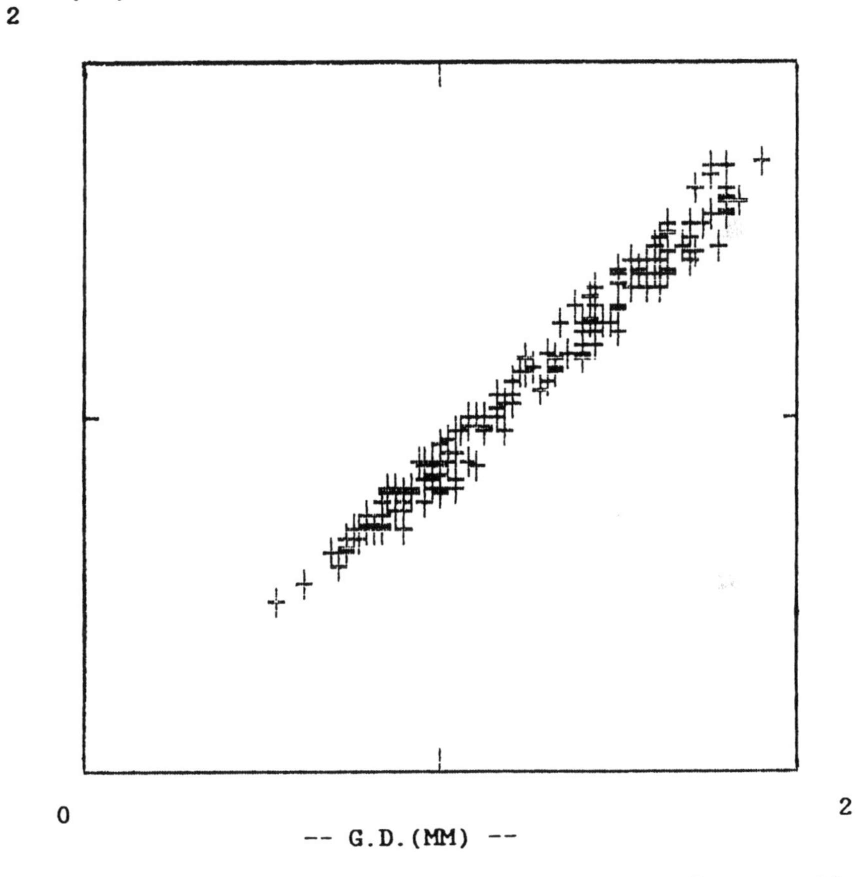

Figure 1. Sample scatter diagram of shell measurements (greater diameter and lesser diameter) in mm of *Elphidium crispum* (Linne'), from author's collection from Mount Soledad (Pacific Beach, CA) (compare with Nicol, 1944).

LATITUDE RANGE IS = 34 TO 36.5
LONGITUDE RANGE IS = -115.5 TO -113.5
DEGREES / SEGMENT = 6

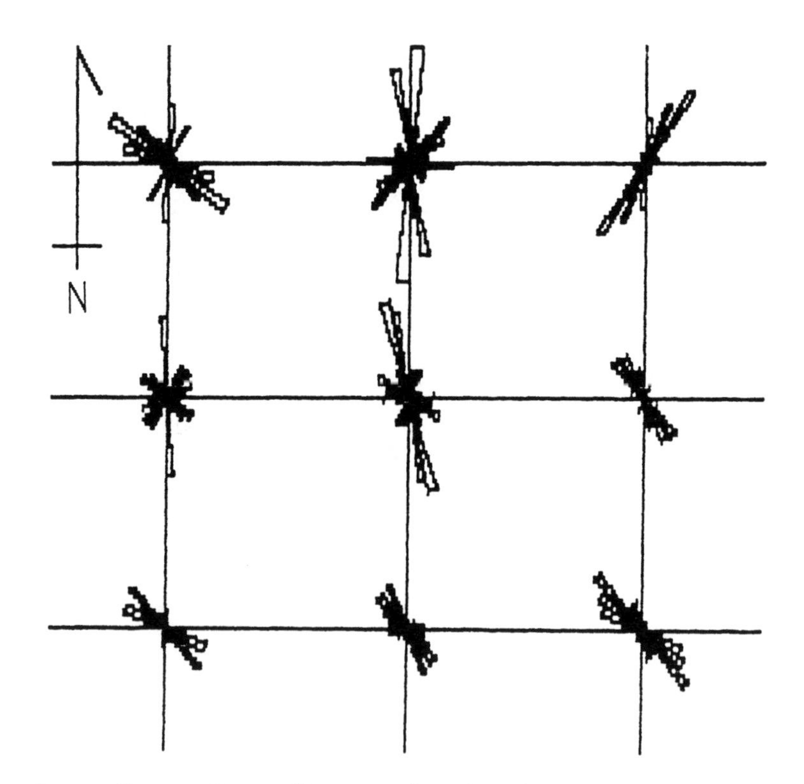

Figure 2. Sample rose diagram for test lineament length and orientation data; geographic area has been divided into 3x3 cell matrix to display subregional trends.

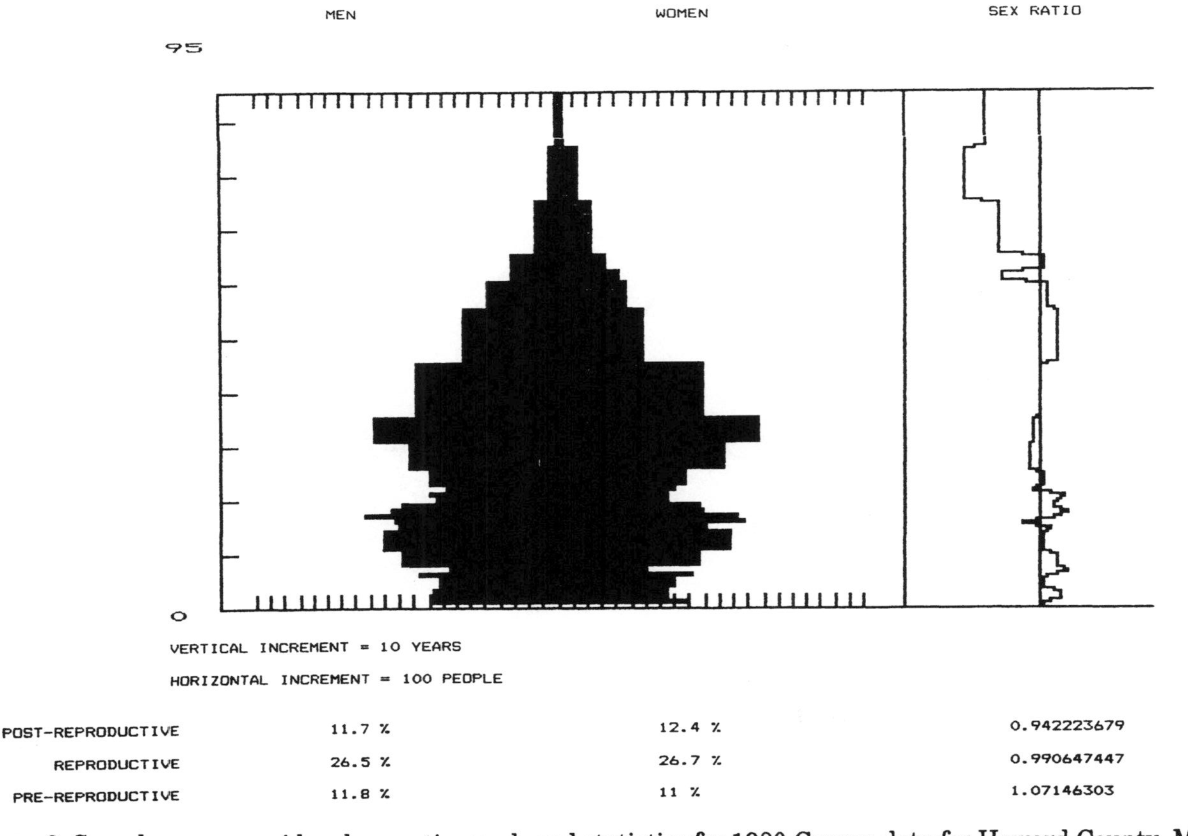

Figure 3. Sample age pyramid and sex ratio graph and statistics for 1980 Census data for Howard County, MD.

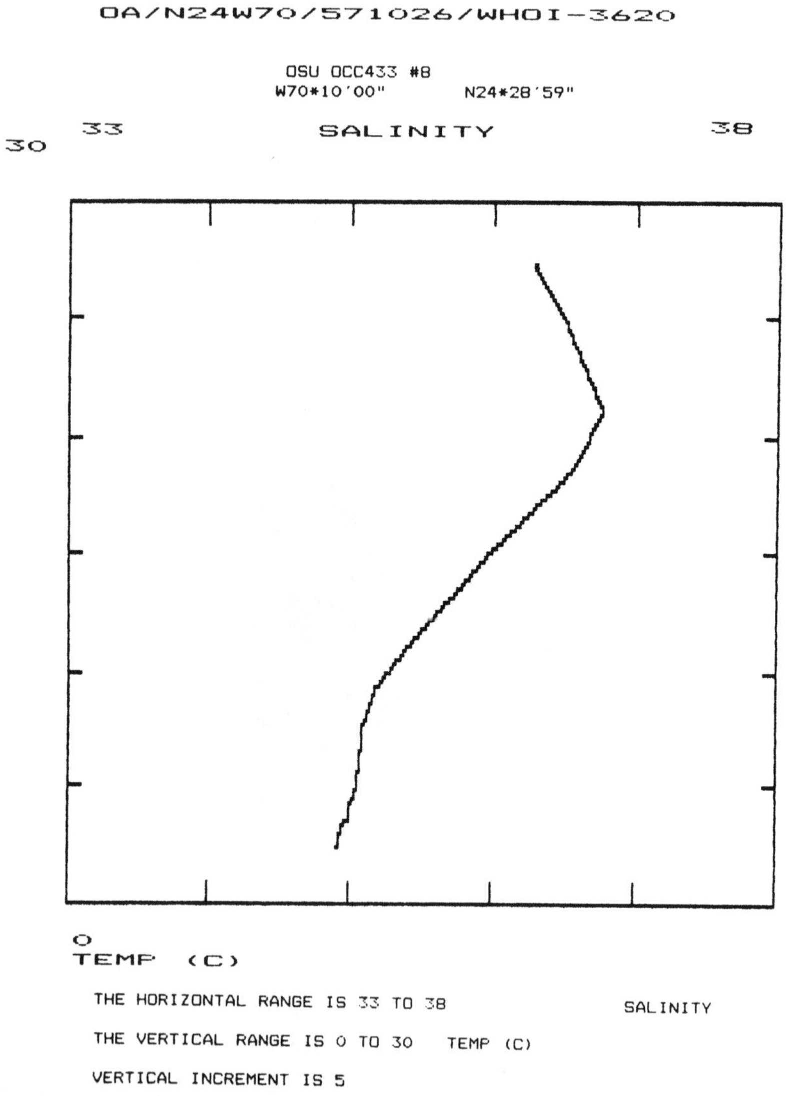

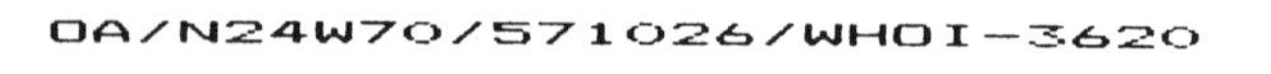

OA/N24W70/571026/WHOI-3620
OSU OCC433 #8
W70*10'00"
N24*28'59"
33
SALINITY
38
30
0
TEMP (C)
THE HORIZONTAL RANGE IS 33 TO 38
SALINITY
THE VERTICAL RANGE IS 0 TO 30 TEMP (C)
VERTICAL INCREMENT IS 5

Figure 4. Sample Temperature-Salinity analysis of hydrographic data from classroom data set showing three water masses; data collected from Atlantic Ocean off Canary Islands.

```
HAND-DIGITIZED TEST DATA

LATITUDE RANGE IS  = 34 TO 36.5
LONGITUDE RANGE IS  = -115.5 TO -113.5

CELL = 1, 1

DEGREES / SEGMENT = 6

AZIMUTHS                  STRIKE-FREQUENCY WEIGHTED BY LENGTH(KM)

 13 - 19                  0
 19 - 25                  0
 25 - 31                  0
 31 - 37                  0
 37 - 43                  0
 43 - 49                  0
 49 - 55                  4              **
 55 - 61                  0
 61 - 67                  0
 67 - 73                  0
 73 - 79                  0
 79 - 85                  0
 85 - 91                  0
 91 - 97                  14             **********
 97 - 103                 0
103 - 109                 0
109 - 115                 10             *******
115 - 121                 40             *****************************
121 - 127                 23             *****************
127 - 133                 24             *****************
133 - 139                 18             *************
139 - 145                 47             ***********************************
145 - 151                 0
151 - 157                 14             **********
157 - 163                 9              ******
163 - 169                 4              **
169 - 175                 11             ********
175 - 181                 6              ****
181 - 187                 18             *************
187 - 193                 5              ***

LINEARS =                 28

PERCENT OF TOTAL =        9.06148868

MINIMUM LENGTH(KM) = 1.8
```

Figure 5. Sample statistics produced on printer; see NW cell of Figure 2.

HAND-DIGITIZED TEST DATA

VARIABLE # -- ID SAMPLE	0 AZIMUTH	1 LENGTH(KM)	2 XMID(W)	3 YMID(N)
1 -- N43W	317	10.5	-115.089	34.712
2 -- N37E	37	10.5	-114.967	35.006
3 -- N06W	354	18.7	-115.013	35.171
4 -- N44E	44	8.2	-114.974	35.225
5 -- N35W	325	23.8	-113.726	35.374
6 -- N40W	320	35.1	-113.943	35.61
7 -- N50W	310	7.02	-114.089	35.77
8 -- N32W	328	7.02	-114.14	35.823
9 -- N18E	18	24.6	-114.115	36
10 -- N34E	34	23.4	-114	36.193
11 -- N52W	308	23.4	-114.013	34.524
12 -- N64W	296	11.7	-113.834	34.374
13 -- N63W	297	21.1	-113.752	34.352
14 -- N08W	352	11.1	-113.752	34.524
15 -- N24W	336	8.2	-113.72	34.412
16 -- N73W	287	3.5	-113.682	34.369
17 -- N61W	299	28.1	-115.388	34.118
18 -- N39W	321	29.8	-114.885	34.283
19 -- N56W	304	8.2	-114.924	34.278
20 -- N11W	349	11.1	-114.847	34.176
21 -- N49W	311	15.2	-115.405	36.417
22 -- N53W	307	28.1	-115.382	36.444
23 -- N67W	293	42.1	-115.032	36.278
24 -- N02.5E	2.5	17.6	-114.529	36.471
25 -- N37.5W	322.5	9.9	-114.171	35.941
26 -- N62W	298	9.9	-115.433	35.936
27 -- N80.5E	80.5	11.7	-114.191	35.567
28 -- N14W	346	17.6	-114.35	35.283
29 -- N36.5W	323.5	17.6	-114.134	34.064
30 -- N26E	26	10.5	-114.159	34.267
31 -- N36E	36	11.7	-113.847	35.749
32 -- N41E	41	3.5	-113.656	35.727
33 -- N0E	0	11.7	-113.592	35.855
34 -- N13E	13	9.4	-113.586	35.952
35 -- N23E	23	9.4	-113.675	36.043
36 -- N9.5E	9.5	8.2	-113.649	36.118
37 -- N35E	35	7.6	-113.796	36.053
38 -- N34E	34	5.9	-113.821	36.064
39 -- N19E	19	11.7	-113.879	36.037
40 -- N33E	33	23.4	-113.879	36.091
41 -- N11W	349	7	-113.649	34.936
42 -- N18W	342	8.2	-113.719	35.064
43 -- N13W	347	11.1	-113.713	34.989
44 -- N07W	353	14	-113.802	35.064
45 -- N00W	0	9.4	-113.821	35.011
46 -- N00E	0	9.4	-113.815	34.497
47 -- N21W	339	6.4	-113.828	34.572
48 -- N15W	345	12.3	-113.93	34.738
49 -- N36W	324	24.6	-113.98	34.219
50 -- N00E	0	9.4	-113.753	35.942

Figure 6. Sample annotated data matrix where number of samples (OCC) > 50 and number of variables (NV) < 6, so that samples are printed 50/page (see Table 1).

Table 1. Common record structure for data files (PEG programs use upwards compatible structure)

Record	Variable	Description
1	NV, OCC	Array size parameter values: NV = number of variables OCC = number of samples
2	ID$	≤256 character string for identification
3	VNAME$(NV)	Array of variable names
4	OCNAME$(OCC)	Array of sample names
5	ARRAY(NV,OCC)	Array of data values

These programs use a common data structure for data records recorded on a disk file using the Apple DOS 3.3 system (Table 1). This allows almost any program to read almost any data file; the user must determine if the analysis is appropriate. Within each program, data arrays are sized dynamically and labeled by parameters read at the start of the file.

These programs also use a common design and resulting program structure (Fig. 7). The development of each program normally reused many code modules from one or more previous programs (Table 2).

CONCLUSIONS

Reusing code makes it easy to develop new programs to fit new applications or new requirements. The reuse of tested code increases software reliability while at the same time increasing developer and programmer productivity, thus reducing cost. The author has determined that new programs can be developed, coded, tested, and placed in operation in as little as one working day by reusing selected modules of previously developed, well-tested code.

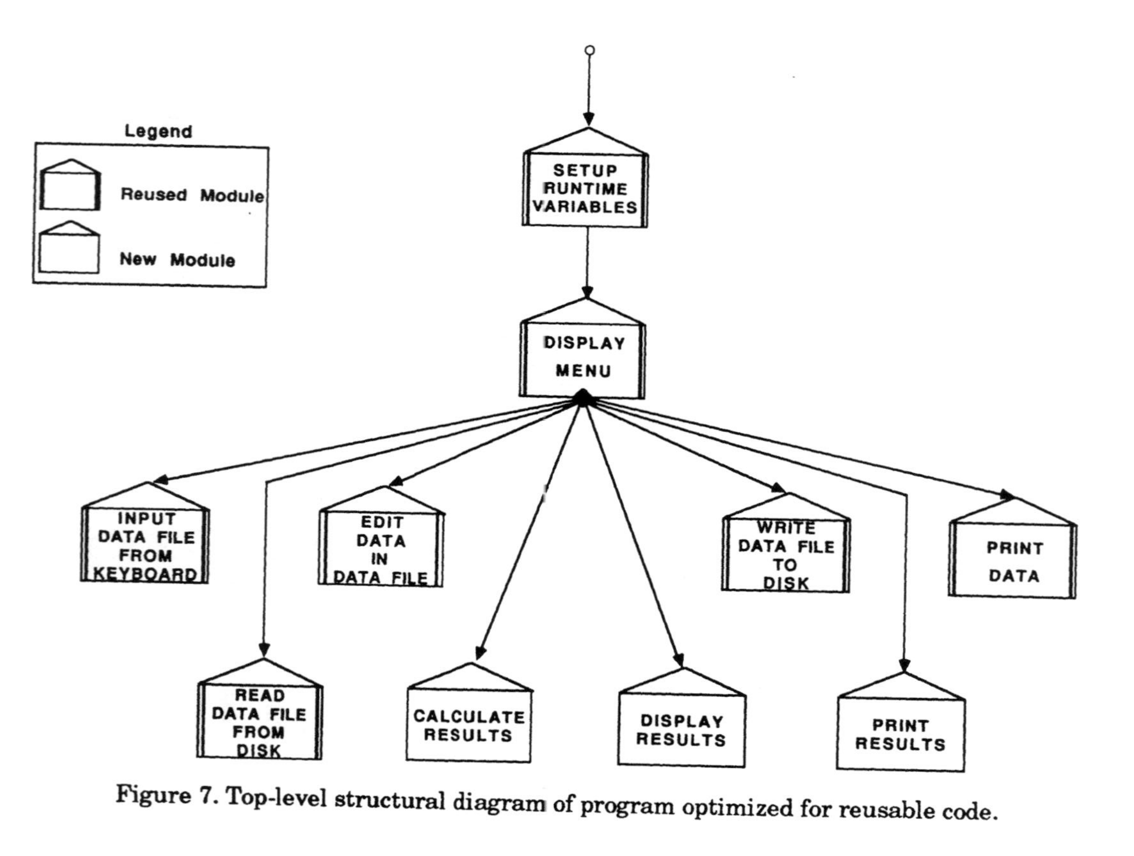

Figure 7. Top-level structural diagram of program optimized for reusable code.

Table 2. Code Reuse History

	1983		1984	1985						1986		1987		1988		1989	
	1	2	3	4	5	6	7	8	9	10	11	12	13	14	15	16	17
	Time Series Plot	XY Plot (V.1)	Calib. Pulse Analy.	Data Mng.	XY-Sc Plot (V.2)	XY -LP Plot	N (XY) Plot	Map Data Mng.	Map Plot	Rose Plot	Age Plot	Basic Data Desc.	RS Plot	Hydro Plot	Profile Plot	PEG Data Mng.	PEG Strat Plot1
Header	New	Mod	Mod	Mod	Mod	Mod	Mod	Mod	Mod	Mod	Mod	Mod	Mod	Mod	Mod	Mod	Mod
Display Plot	New	#			#	#	Mod		Mod	New	New		#	Mod	#		#
Print Graph	New	#			#	New	Mod		#	#	#		#	#	#		New
Create Data File	New	#	#	#				#								Mod	
Edit Data in File	New	#	#	#				#								Mod	
Write Disk File	New	#	#	#				#								Mod	
Read Disk File	New	#	#	#	#	#	#	#	Mod	#	#	#	#	Mod	#	Mod	#
Do Special Work	New	#	Mod		#	#	#		New	#	#	#	#	New	New	New	New
Setup Work	New	#	#	#	#	#	Mod	#	Mod	New	#	#	#	New	New	#	Mod
Main Loop	New	#	#	#	#	#	#	#	#	#	#	#	#	#	#	Mod	Mod
Setup Plot	New	#	#		#	#	#		Mod	New	#		#	Mod	#		Mod
Print Data	Stub	New	#	#	#	#	#	#	#	New	#	#	#	#	#	Mod	Mod
Modify Data				New				#								#	
Setup Function	New	#	#	Mod	#	#	#	Mod	Mod	Mod	#	#	#	New	New	Mod	Mod

 # - Used "as is" or with minor modifications
Mod - Major modifications
New - Extensive rewriting or new code

REFERENCES

Gunther, F. J., 1982, Statistics of thematic maps for users of Landsat data and Geographic Information Systems: Computer Sciences Corporation, CSC/TM-82/6052, Silver Spring, Maryland, 50 p.

Gunther, F. J., 1984, A comparison of Landsat-5 thematic mapper calibration pulse integration methods: Computer Sciences Corporation, CSC/TM-84/6097, Silver Spring, Maryland, 58 p.

Gunther, F. J., 1986, Advanced statistical analysis of lineament and orientation data with a personal computer: Geol. Soc. America, Northeast Section, Abstracts with Programs, v. 18, no. 4, p. 290–291.

Gunther, F. J., 1987, Using a microcomputer to plot Landsat and SPOT scene boundaries as a purchasing aid: Proc. Pecora XI Symposium, Sioux Falls, South Dakota, p. 367–375.

Nicol, D., 1944, New West Coast American species of the foraminiferal genus *Elphidium*: Jour. Paleontology, v. 18, no. 2, p. 172–185.

A PC STATISTICAL PACKAGE FOR FIELD ACQUISITION AND ANALYSIS OF TWO-DIMENSIONAL ORIENTATION DATA

N. I. Fisher, *CSIRO, Sydney , Australia*
C. McA. Powell, *University of Western Australia, Perth, Australia*
A. Gelin, *Macquarie University, Sydney, Australia*
and
D. McP. Duncan, *Department of Mines, Tasmania, Australia*

ABSTRACT

We describe a newly developed PC statistical package which, in conjunction with a suitably designed laptop or rugged field computer, can be used by practicing geoscientists collecting orientation data in the field. Potential users include structural geologists, field mappers, sedimentologists, and any scientists gathering orientation data. This paper illustrates how the package might be used in practice, in summarizing individual sets of measurements and comparing and combining the summary values from the various sets.

DESCRIPTION OF PACKAGE

The package uses the most up-to-date statistical methods available for analyzing single sets of orientations, and for comparing and combining several sets of measurements based on the large-sample methods devised by Fisher and Lewis (1983) and Watson (1983), and the bootstrap techniques of Fisher and Hall (1989, 1990), and an unpublished report "New statistical methods for directional data II: bootstrap confidence regions for mean directions"). It offers the systematic framework for handling such data described by Fisher and Powell (1989). It is IBM-compatible, MS DOS-based and menu-driven. It can be used in a PC in a laboratory, where the extended graphics capabilities of a PC can be exploited. However, we envisage that its real value will be in its application in the field, in conjunction with a sturdy field-portable PC. The geoscientist will be able to log orientation data and other field information directly into the computer, process the data immediately to decide whether enough have been collected to form reliable estimates, compare measurements from different sites to identify trends, and so on. At the end of a day, or a field trip, the data and results then can be transferred directly to a database or a larger IBM-compatible computer and plotter.

A typical field operation involves

(a) entry of general information about the field site;

(b) entry of paleocurrent (or other structural) data, and on-the-spot analysis to check that all data satisfy reliability criteria, and that the total data set meets desired confidence levels; and

(c) comparison and synthesis with other data sets.

The procedure is illustrated in the next section.

EXAMPLE

Here we show how the package might be used in practice. The data would be collected in the field together with other relevant in-situ information, and stored in a file XBGR26A, the contents of which would appear as printed in Table 1.

Using the statistics described in Fisher and Powell (1989), this data set can be summarized by line 1 of Table 2. Lines 2, 3, and 4 are similar data sets from adjacent localities. At this point, we may wish to compare the mean directions of the localities and if appropriate, pool them to form an overall estimate of the mean direction.

Using the large-sample test given in Fisher and Powell (1989), we determine that localities 1, 2, and 3 can be taken to be drawn from

Table 1. Field data

File: XBGR26A

Description: Red Mans Bluff Sandstone, Basal Unit, Cross-Bedding Fluvial, Medium-to Coarse-Grained Quartzarenite, Horsham, 1:100 000 Grid Ref. 226158, April 13, 1987. Pebbles to 41 mm long dimension, coset approximately 20 m thick.

Data (in Dip/Dip Azimuth Format)

	Regional Bedding	Cross-Bedding Measured	Restored	Set Thickness (cm)
1	24/021	52/026	28/029	17
2	(mean of	44/355	24/333	15
3	4 measurements)	49/346	32/326	26
4	"	38/324	31/289	19
5	"	44/359	23/340	17
6	"	48/346	31/325	29
7	"	38/000	18/334	24
8	"	50/005	28/354	20
9	"	40/005	18/346	35
10	"	34/320	30/281	30
11	"	42/000	21/339	24
12	"	51/335	22/330	8
13	"	41/011	18/359	35
14	"	36/359	16/328	30
15	"	42/034	19/048	15
16	"	55/018	31/016	25
17	"	46/005	24/351	35
18	"	49/022	25/023	35
19	"	42/010	19/358	19
20	"	48/015	24/010	16
21	"	41/012	18/001	8
22	"	34/021	10/021	150
23	"	39/348	22/317	10
24	"	30/350	15/300	53
25	"	32/345	18/300	35
26	"	60/026	36/028	45
27	"	26/341	17/279	267
28	"	41/358	21/335	70
29	"	36/041	15/070	55
30	"	46/359	25/341	17

Table 2. Data summary

	Filename	No of Vectors	Median	Vector-mean Azimuth	Mean Resultant Length	Circular Standard Error	95% Confidence Interval	Quartiles	
		n	$\tilde{\theta}$	$\hat{\mu}$	$\bar{R}$	$\hat{\sigma}$	α^0_{95}	$\theta_{.25}$	$\theta_{.75}$
1	XBGR26A	30	340.5	344.65	0.8228	0.1159	13.13	325.5	010.1
2	XBGR27A	34	339.0	342.61	0.8064	0.1155	13.09	318.6	016.6
3	XBGR07A	30	357.0	352.47	0.8284	0.1158	13.12	324.9	017.4
4	XBGR13A	30	076.5	072.17	0.7698	0.1330	15.11	45.0	102.7

populations with the same mean direction; the mean direction of locality 4 differs significantly from the others. It therefore is appropriate to form a pooled estimate of the common mean direction for the first three localities.

Another example is given in Taylor and Mayer (1990, table 1) where paleocurrents from the fluvial Upper Devonian Worange Point Formation, NSW, are analyzed and combined to form higher order pooled estimates of the common mean direction.

The advantages of doing these statistical tests in the field are:

(1) The data can be checked to eliminate any measurements which fall outside tolerable limits (e.g. foreset dips < 8° or > 40°). New data can be measured to replace rejected data.
(2) The data can be checked to see if acceptable precision (expressed partly by confidence interval) has been achieved - although note that large amounts of dispersion in the data are characteristic of certain populations, and may result in wide confidence limits.
(3) Checks can be made as to whether the mean directions are similar or different from previously measured data, thus improving knowledge about the changing paleocurrent patterns as the data are being measured.

DISCUSSION

Three of the localities (1, 2, and 3) have paleocurrent data which satisfy statistical tests that they have drawn from populations with a common mean direction, and one locality (4) contains data whose mean orientation differs significantly from the others. Nonetheless, all four localities are in the same geological unit (the basal pebbly band of the Red Mans Bluff Sandstone), and there may be geological reasons why one might want to use information from all four localities to calculate a grand mean for the entire unit. For instance, each locality represents a single estimate of the mean orientation of the paleostreams which formed the crossbeds. If we wish to look at the mean orientation of all such estimates of paleostream orientation across an area, or in a mappable geological unit, we are entitled to take the estimate of mean paleocurrent flow at each locality, and, using appropriate weighting factors such as the areal distribution or the percentage of the geological unit which the paleocurrent information represents, we could combine the vector-mean azimuths into a higher order estimate of the mean paleocurrent direction for the unit (Potter and Pettijohn, 1977, p. 383-384).

The important aspect treated in this paper concerns statistical tests that allow us to determine whether the orientation data from any two or more localities can be regarded as being drawn from populations with a common mean direction, and, if they cannot, prompts us to ask the question of what the significance of that observation is. The statistical tests do not prevent us from amalgamating or merging data, if it can be argued on geological grounds that such merging should occur. For example, the crossbeds used in this example are a sedimentary structure of rank 5 (Table 3, after Miall, 1974), and form the basic direction structure from which geologists try to answer questions about paleoslope requiring information two or three rank orders higher.

Information about paleoslope directions can only be acquired properly if appropriate sampling strategies are adopted, and field information is collated and weighted according to the directional question being asked. Thus, whereas 30 crossbed measurements in one coset representing a single paleostream can give an accurate estimate of the flow direction in that paleostream, how accurate is the estimate of regional paleoslope given that one paleostream flow direction? A more accurate estimate of the paleoslope would be given by taking a few crossbed measurements in several paleostreams, so that the variability of paleochannel orientation could be smoothed out.

A full discussion of the strategies and procedures involved in paleocurrent sampling is beyond the scope of this paper, and is mentioned here as a caution to guide sensible use of the statistics we present herein.

Table 3. Rank hierarchy in fluvial systems (after Miall 1974)

Rank	Type of structure	Source of observations
1	Entire drainage systems	Regional lithological, facies, paleocurrent and isopach maps.
2	Meander belts of individual rivers.	Detailed lithological, facies, paleocurrent and isopach maps.
3	Major channel reaches within meander belts.	Detailed lithological, facies, paleocurrent and isopach maps.
4	Minor subsidiary channels, transverse, lateral and point bars.	Channel axes, epsilon crossbeds.
5	Structures within bars.	Crossbedding, trough axes.
6	Structures superimposed on rank 5 structures	Ripple marks, small-scale cross-lamination, pebble imbrication.

APPLICATION TO STRUCTURAL ANALYSIS

The statistical methods illustrated in this paper have much broader application than paleocurrent analyses. The methods can be used for any structural analysis where linear or directional data are involved. Two examples are the analysis of joints and aerial-photograph lineaments, and the analysis of fold axes and lineations in ductily deformed terranes.

The analysis of joint distributions and orientations has great economic significance in the coal-bearing Sydney Basin, and has been the subject of several substantial investigations in the past decade (e.g. Shepherd and Huntington, 1981; Cudahy and Creasey, 1986). Our statistics enable these joints distributions to be analyzed in an objective, quantifiable way, and, by use of the summary statistical tables, information gathered later can be added to existing information without the need to reprocess the entire raw data set.

Statistical analysis of fracture and lineation data in folded terranes is likely to become increasingly important in the mining industry, especially as the open-cut phase of gold mining leads to the deeper level underground mines where ore shoots usually follow fracture or fold directions. At present, there are few statistical packages developed specifically for quantifying such mining problems, and we intend to extend our present analysis to three dimensions.

ACKNOWLEDGMENTS

Development of this PC package was supported by a CSIRO-Macquarie University Research Grant and the Australian Research

Council. Extensive field work which underpins the data, and led to clarification of the problems to be addressed, has been supported by Macquarie University and ARGS research grants.

REFERENCES

Cudahy, T.J., and Creasey, J.W., 1986, The role of basement structure in controlling structural and sedimentary pattern in the southern Sydney Basin: CSIRO Inst. Energy and Earth Resources, Invest. Rep. No. 1628R, 3 vols.

Fisher, N.I., and Hall, P., 1989, Bootstrap confidence regions for directional data: Jour. Am. Statist. Assoc., v. 84, no. 408, p. 996-1002.

Fisher, N.I., and Hall, P., 1990, New statistical methods for directional data I. Bootstrap comparison of mean directions and the fold test in paleomagnetism: Geophys. Jour. Royal Astr. Soc., v. 101, p. 305-313.

Fisher, N.I., and Lewis, T., 1983, Estimating the common mean direction of several circular or spherical distributions with differing dispersions: Biometrika, 70, No. 2, p. 333-341; Correction, v. 71, no. 3, 1984, p. 655.

Fisher, N.I., and Powell, C. McA., 1989, Statistical analysis of two-dimensional paleocurrent data: methods and examples: Aust. Jour. Earth Sci., v. 36, p. 91-107.

Miall, A.D., 1974, Paleocurrent analysis of alluvial sediments: a discussion of directional variance and vector magnitude: Jour. Sed. Pet., v. 44, no. 4, p. 1174-1185.

Potter, P.E., and Pettijohn, F.J., 1977, Paleocurrents and basin analysis (2nd ed.): Springer-Verlag, Heidelberg, 425 p.

Shepherd, J., and Huntington, J.F., 1981, Geological fracture mapping in coalfields and the stress fields of the Sydney Basin: Jour. Geol. Soc. Aust., v. 28, p. 299-309.

Taylor, G., and Mayer, W., 1990, Depositional environments and paleogeography of the Worange Point Formation, New South Wales: Aust. Jour. Earth Sci., v. 37, p. 227-339.

Watson, G.S., 1983, Statistics on spheres in Univ. Arkansas Lecture Notes in the Mathematical Sciences, Vol. 6: John Wiley Interscience, New York, 238 p.

A SIMPLE METHOD FOR THE COMPARISON OF ADJACENT POINTS ON THEMATIC MAPS

James C. Brower
Syracuse University, Syracuse, New York, USA

and

Daniel F. Merriam
Kansas Geological Survey, University of Kansas, Lawrence, Kansas, USA

ABSTRACT

A simple method is outlined for comparing adjacent grid points that have been measured for a series of maps. It is generalized and can be calculated for different types of maps where the original data are continuous or discrete. Either original or standardized data can provide the input information; standardization expresses the various maps in the same units. Next, similarities or differences are computed for all adjacent points from east to west and from north to south on the grid. The coefficients computed include correlation coefficients, Euclidean distances, and Mahlanobis distances; however, other statistics could be employed where appropriate. The coefficients then are plotted on the grid and contoured to depict the distribution of similarities and differences. Various patterns of similarities and differences between the points are shown by different coefficients and standardizations which can be related

Use of Microcomputers in Geology, Edited by D.F. Merriam and H. Kürzl, Plenum Press, New York, 1992

to geologic features underlying the original data. The maps of the point-to-point comparisons are suitable for subsequent study with other methods such as trend surfaces, filtering, or Fourier analysis. The example given is based on five structure contour maps from the Paleozoic of Kansas.

INTRODUCTION

In geology the analysis of spatial data is a major aspect of many studies (Merriam and Jewett, 1988; Brower and Merriam, 1990; Herzfeld and Merriam, 1991). The relationship of the data points to each other is important also as are the assumptions made in applying statistical techniques to the data. Therefore, understanding these interrelations can affect the outcome and the interpretation. Consequently any method that can show and evaluate these interrelationships can aid in the investigation of mapped data.

The approach proposed here is one such method. We have taken a gridded data set that was compiled for a spatial data integration and comparison study to test the validity of this approach (see Brower and Merriam, 1990; Table 1). Examination of the interrelations of adjacent data points for five structure contour maps was made to determine what if anything could be learned. The comparison of the neighboring data points was based on standardized and nonstandardized data; the distribution of the similarities and differences between the adjacent data points shows the spatial patterns of similarities and differences.

Interpretation of the maps is not straightforward and requires some knowledge and background in the geology and data set of the area. Nevertheless the results are interesting and give the investigator new insight into the interrelations of the data and how they might affect the results from other analyses.

A study by Merriam and Sondergard (1989) explored the use of a Reliability Index (RI) to determine where pairwise comparisons of maps were in high correspondence and thus had good predictability. The effect of this analysis was to look at the relationships between variables (the maps); whereas this study determines the relationships between samples (the data points). Merriam and Sondergard showed the RI indicated that matches were high (good) over most of the geological maps compared except for local anomalous areas, but that there was little correspondence between different types of thematic maps (eg. geological/geophysical, geological/topographic, and geophysical/topographic).

Table 1. List of map data. Units consist of elevations in feet below sealevel

MAPS					
Sample Number	Lansing	Kansas City	Mississippian	Arbuckle	Precambrian
1	770	1160	1395	2100	2500
2	690	1040	1310	1760	2300
3	590	890	1150	1600	2100
4	480	800	1005	1290	2010
5	350	705	865	1210	1750
6	290	670	1000	1300	1750
7	350	310	950	1400	1000
8	825	1185	1490	2090	2710
9	700	1100	1420	1800	2520
10	600	980	1210	1670	2240
11	475	795	1050	1400	1900
12	300	705	950	1290	1770
13	350	680	950	1400	2000
14	435	790	1385	1720	2450
15	880	1250	1605	2180	2750
16	800	1185	1440	1965	2590
17	600	1025	1300	1770	2375
18	560	850	1080	1300	2000
19	440	735	990	1410	1950
20	460	800	1200	1300	1500
21	405	750	1350	1770	2405
22	930	1340	1600	2100	2840
23	820	1205	1525	2050	2695
24	695	1030	1380	1770	2430
25	430	805	1100	1300	2000
26	380	740	1010	1400	1950
27	250	500	1000	1500	1300
28	420	725	1325	1820	2400
29	940	1395	1700	2250	2980
30	840	1230	1595	2100	2740
31	750	1120	1460	1850	2590
32	620	1005	1300	1750	2500
33	475	825	1125	1300	2350
34	450	550	1200	1810	1980
35	390	800	1320	1825	2380
36	1000	1405	1690	2150	2930
37	850	1270	1610	2100	2800
38	725	1160	1480	1950	2615
39	550	1000	1315	1500	2300
40	180	700	1100	1080	1800
41	460	900	1250	1700	2525
42	200	835	1315	1800	2485
43	1040	1490	1800	2260	3100
44	850	1310	1610	2000	2900
45	675	1205	1505	1900	2640
46	590	1100	1410	1700	2420
47	550	900	1300	1700	2500
48	500	1020	1410	1850	2510
49	320	810	1305	1720	2400

The normal assumption in any geological study is that the data points are independent, but geostatisticians have shown in recent years that this assumption usually is not valid (eg. Journel and Huijbregts, 1978). This study confirms that adjacent data points may be highly correlated especially in areas where there is a high degree of coincidence of structural features on the maps. This is the geological situation with this data set from Kansas where structures tend to persist and normally become better defined and sharper with increasing depth (Merriam, 1963).

DATA

Five structure contour maps from eastern Kansas were selected to illustrate our method of point-to-point comparisons; these include the top of Precambrian (Cole, 1962), top of the Ordovician Arbuckle Group (Merriam and Smith, 1961), top of Mississippian (Merriam, 1960), base of the Pennsylvanian Kansas City Group (Watney, 1978), and top of the Pennsylvanian Lansing Group (Merriam, Winchell, and Atkinson, 1958) (Figs. 1A-E). The original units are elevations in feet below sealevel. Forty-nine points, located at six-mile intervals on a 7 by 7 square grid, were measured on each map to generate the test data (Fig.1F; Table 1).

The data set is limited in several respects. (1) Inasmuch as the five maps are structure contours, all are of the same basic type. For most applications, one would want to compare different types of maps, such as structure, topography, gravity, magnetics, lithofacies, biofacies, paleocurrents, etc. (2) The five maps of this data set are all measured at the same points, the 7 by 7 grid in this instance. This may not be the situation as for example with subsurface data, where all wells do not reach all of the units that would be involved in the comparisons. Here, gridding would be necessary because the grid points would provide the input data.

NUMERICAL DATA

The intent of this study is to depict point-to-point similarities and differences with respect to all of the maps. These have an obvious significance for reconstructing the geologic history of an area or exploring for economic deposits of petroleum or minerals. The method has the virtue of simplicity if nothing else. The basic concept of studying the spatial relationship between similar to the idea of geostatistics (the theory of regionalized variables).

Figure 1. Structure contour maps from eastern Kansas. Units are elevations in feet below mean sealevel. A. Lansing; B. Kansas City; C. Mississippian; D. Arbuckle; E. Precambrian; F. Map of 49 grid points.

Similarity or distance coefficients are calculated for all adjacent pairs of points from east to west and from north to south on the grid. For the first row of the grid, the coefficients for the east-west comparisons would involve points 1 and 2, 2 and 3, 3 and 4, 4 and 5, 5 and 6, 6 and 7. The analogous coefficients from north to south for the seventh and last column of the grid would be calculated for points 7 and 14, 14 and 21, 21 and 26, 26 and 35, 35 and 42, 42 and 49 (see Fig. 1F). The coefficients then are plotted on the grid and the patterns of similarities or differences are revealed by the contours. Note that this scheme preserves most of the original dimensions of the maps except for "bites" at the corners and along the edges.

Comparison of the points based on standardized or unstandardized data with various coefficients shows different aspects of similarities and differences. Analyses were determined for the original structure contours in feet below sealevel and data where each map was standardized by Z-Scores or standard deviation scores. This form of standardization expresses the maps in the same units so that each map will contribute equally to the coefficients computed for the parts of points. Because of this transformation, the high and low areas of a single maps are associated with low negative and high positive Z-scores, listed in the same order. The coefficients selected for the point-to-point comparisons consist of Pearson-product-moment correlations, Euclidean distances, and Mahlanobis coefficients correct for redundant information between the maps because the distances between the points, say i and j, are weighted inversely with respect to the pooled covariances for the variables or maps. The analyses are not limited to the coefficients employed here, and others, for example Manhattan metrics, could be treated if desired.

Although we have not done so, the original coefficients could be replaced with their significance or probability levels. For example, one might contour the probabilities that the correlation or distance coefficients differ from nil. Positive and negative correlations would be indicated by the corresponding signs. Henderson and Heron (1977) and Raup and Crick (1979) adopted this approach in working with probablistic similarities for paleoecological and biogeographical information.

The maps of the point-to-point comparisons can be used for later analyses with techniques such as trend surfaces, filtering, or Fourier analysis.

In the final step of the study, the resulting contour maps of the point-to-point similarities and differences are compared with unweighted-pair-group-method cluster analysis of matrices of correlation coefficients and absolute values of correlation coefficients. The input data comprise the vectors of the coefficients comparing the adjacent pairs of points.

RESULTANT MAPS

Figure 2A contains the map of point-to-point Eculidean distances where the data on each map were standardized by Z-scores. The most prominent feature of the map is the elongate northeast-trending feature on the eastern side where the largest distance coefficients, ranging from 1.01 to 1.98, are located. This area corresponds to the Nemaha Anticline (see Fig. 1) as would be expected inasmuch as the most rapid changes take place in this region. It is interesting and important to note that the underlying structure is reflected clearly in the distances between the adjacent points of the grid. Smaller distance values are present in the western part of the map which is relatively featureless in terms of the underlying structure. The "boxy" configuration is not truly realistic but the geologic features in this region do have a northeast grain.

The map based on Euclidean distances and the original elevations in feet below sealevel is pictured in Figure 2B. Here, the distances span a large interval of 19.0 to 241.8 because the data were not standardized. This map closely replicates the previous one. The Nemaha Anticline along the eastern side again is marked by large distances. The large anomaly in the northeastern corner of the map is present on both Euclidean distance maps. In addition, the position of the large distances outlining the Nemaha Anticline is nearly the same in both maps. The main contrast between the two maps is that in the western part where the trends are slightly less prominent and the features are more diverse.

The configuration of the map of the Mahlanobis distances (Fig. 2C) is similar to that of the two Euclidean distance maps except that the variation of the distances is lower. This is probably the result of the standardization introduced into Mahlanobis distances by scaling them inversely with respect to the pooled covariance matrix between the variables of maps (see Sneath and Sokal, 1973). The distances range from 0.43 to 5.85 with the largest values being concentrated along the Nemaha Anticline. The structure in the Salina Basin in the western part of the map shows up as a series of rather subdued features of low magnitudes that are elongated from northeast to northwest similar to maps shown in Figure 2A and to a lesser extent 2B.

The map for the correlation coefficients between the points based on the original data is given in Figure 2D. The Nemaha Anticline dominates

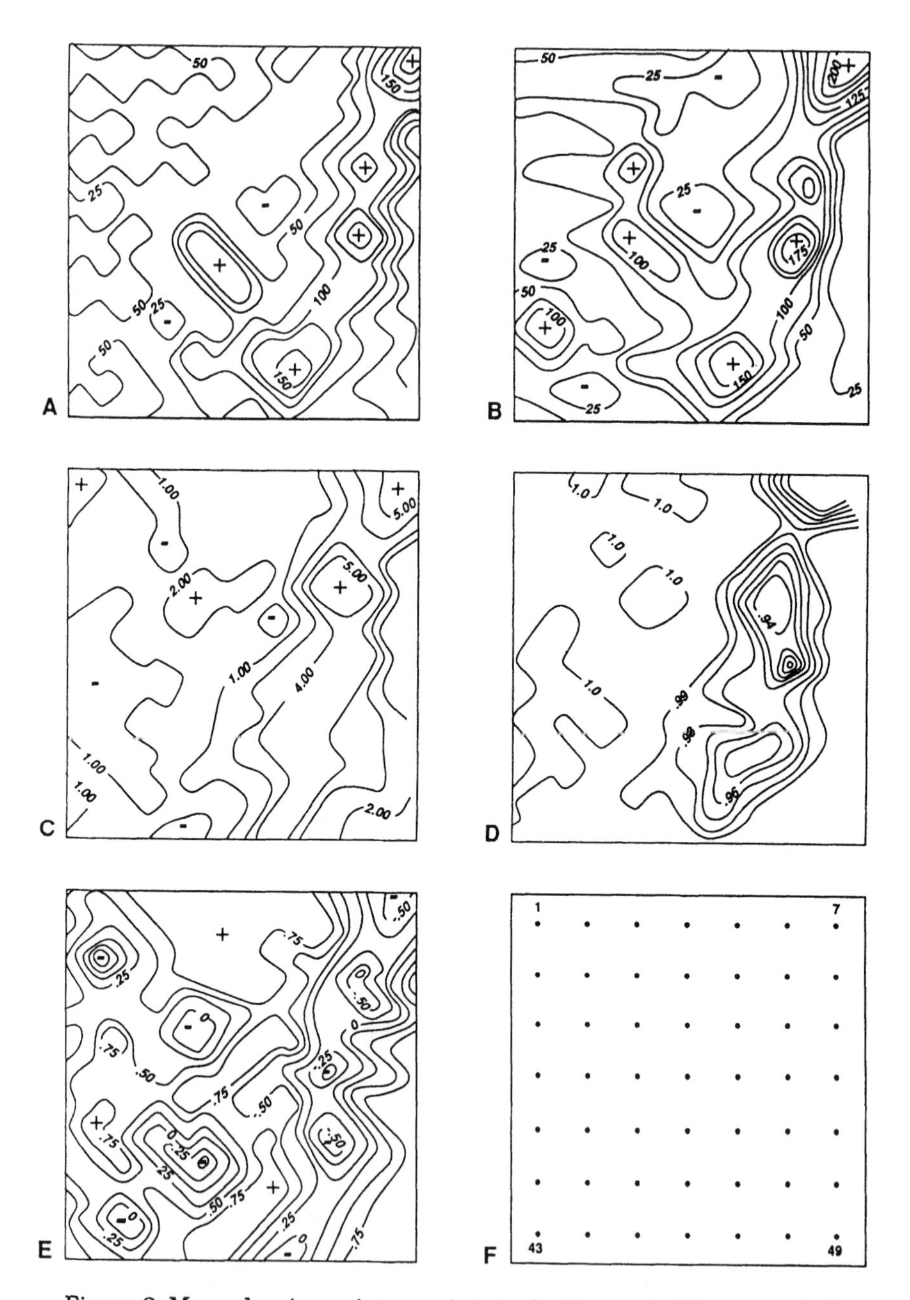

Figure 2. Maps showing point-to-point similarities and differences. A. Euclidean distances for standardized data; B. Euclidean distances for original data; C. Mahlanobis distances for original data; D. Correlations for original data; E. Correlations for standardized data; F. Map of 49 grid points.

the map, and it extends across the eastern side of the map in a northwesterly direction. The contours in the western area are irregular and difficult to characterize although the prevailing orientations are northeast and northwest. Scrutiny of the structure contour maps (Fig. 1) indicates that there are many small-scale structural features in the western area; however, these cannot be detected on the map of the correlation coefficients because of the relatively coarse size of the grid. Visual inspection of the map of the correlations and the structure contour maps makes it obvious that the lower correlations (down to 0.771) are associated with the Nemaha Anticline which has the most structural relief. Conversely, the higher correlations lie in the relatively flat and featureless eastern flank of the Salina Basin where most of the correlations are on the order of 0.0999. The underlying interpretation of this pattern is straightforward. In relatively featureless areas the adjacent grid points exhibit similar values for all of the structure contour maps which produces the large correlations. With more intense structural features, the data can change more rapidly between neighboring points; in addition, the magnitudes of the changes usually are not consistent on all maps. These factors interact and result in lower correlation coefficients between the neighboring points can be related clearly to the patterns on the structure contour maps.

Figure 2E portrays the geographic variation of the correlation coefficients that were calculated from the standardized structure contour maps. The strong "boxy" nature of the configuration is most likely a function of the contour interval and the coarse structure of the grid. Interestingly enough, the crest of the Nemaha Anticline, a strong positive structure, is represented by a series of large negative correlative structure, is represented by a series of large negative correlation coefficients. West of the Nemaha Anticline, the correlations range from positive to negative values which roughly correspond in area to structures, but they may differ in sign. In other words, positive features are associated with negative structures (or synclines) and vice versa.

RELATIONSHIPS BETWEEN THE MAPS

Cluster analysis of the unweighted-pair-group-method type displays the similarities and differences between the maps of the point-to-point relationships (Sneath and Sokal, 1973). The data represent the vectors of coefficients comparing the adjacent points. The dendrogram for the correlation coefficients between the maps yields two distinct groups

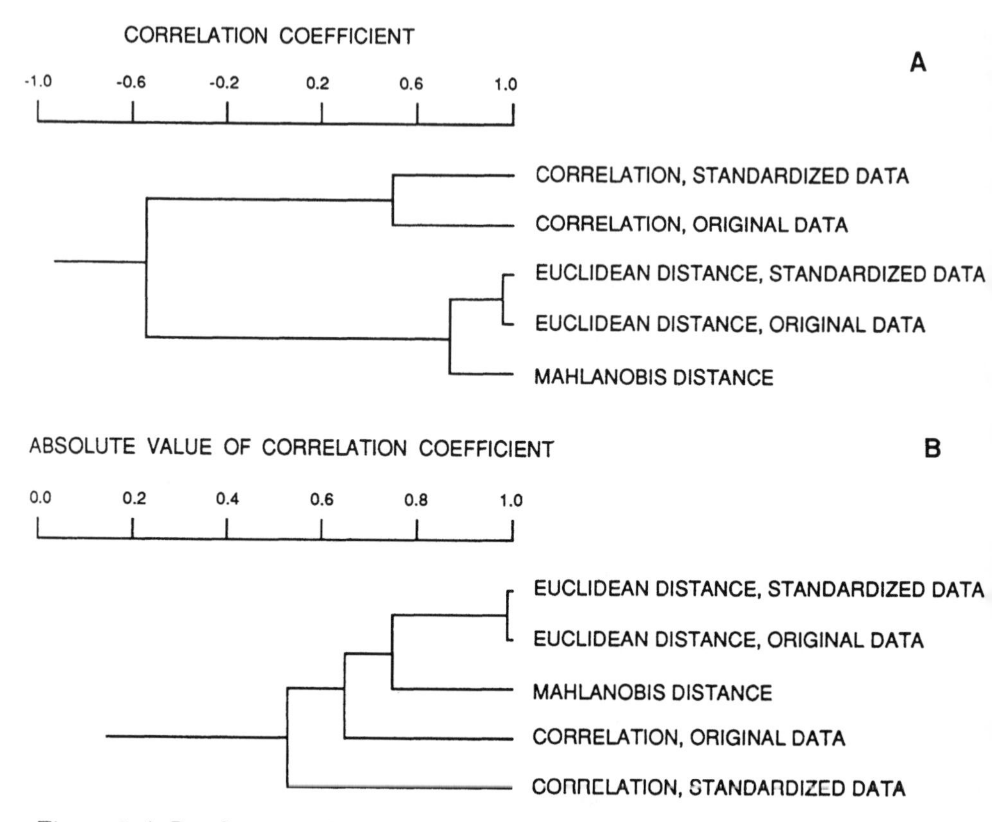

Figure 3. A. Dendrogram of correlation coefficients; B. Absolute value of correlation coefficients.

which are linked at a correlation of -0.562; one cluster contains the three distance maps whereas the other includes the two based on correlations (Fig. 3A). Inspection of Figure 2 demonstrates that the distances and correlations are inversely related. Areas characterized by large distances exhibit comparatively low correlations and vice versa as would be expected (compare Figs. 2A-2C with 2D and 2E). Within the three distance maps, the one for Mahlanobis distance is least similar, mainly because of the different orientations of the patterns of similarities and differences in the western side of the maps. The relatively low similarity between the two maps of the correlation coefficients is dictated by differential variation of the similarity coefficients of the western data points.

The matrix of the absolute values of the correlations should better reflect the underlying similarities of patterns between the five maps of the point-to-point relationships. The dendrogram contains a single rather loose cluster where the two maps of Euclidean distances are nearly identical (Fig. 3B; compare Figs. 2A and 2B). The maps for Mahlanobis distance and correlation coefficients of the original data form the next joiners into the cluster. The latter maps resemble the first two in the vicinity of the Nemaha Anticline but contrasts become apparent in the western part of the maps. The contours on the two Euclidean distance maps are elongated from east to west but those of the Mahlanobis distance and correlation maps are relatively featureless and blocky (compare Figs. 2A and 2B with 2C and 2D). The one for correlation coefficients and standardized data comprises the "odd map out," possibly because the standardization process generates more detail in the western region.

INTERPRETATION

Visual inspection of the five maps of the point-to-point similarities and differences shows that most prominent features on the maps are coincident with major structural features on the structure contour maps. The Nemaha Anticline is the dominant structure extending across eastern Kansas from northeast to southwest. The Salina Basin lies in north-central Kansas west of the Nemaha where it forms a gentle, approximately-symmetrical basin which continues northward into Nebraska. The detailed structure of the Salina Basin is not evident on contour maps constructed from well logs because the data are sparse; these structures are clear only on seismic surveys or other geophysical maps. The eastern flank of the Salina Basin does contain several small structure which generally parallel the Nemaha Anticline, and they can be recognized by workers familiar with Kansas geology. Some of these structures carry through on the five maps although they may be offset slightly.

The map of Figure 2D is the easiest to interpret. The Nemaha is apparent as is the northeast and northwest orientations of the smaller structures of the Salina Basin. On the map for Euclidean distances of the standardized data (Fig. 2A), the Nemaha is certainly obvious, and some of the small structures of the Salina Basin seem to be trending subparallel to the major feature. Where relief is small, note that the data could be contoured differently using the same grid spacing and contour interval.

The association of the strong (large) negative values with positive structural features on the map of the correlation coefficients for the Z-scored data is not readily understandable (Fig. 2E). However, this association is consistent with the interpretation of the geology of the Salina Basin. If the units of the original maps are eliminated by standardization, basic geometry would indicate that the correlation coefficients will either increase or decrease in the region where the axis of a large flexure present on all or most of the original maps is crossed. The maps of Figures 2C-2E suggest also that the structures in the Salina Basin have a northeast or northwest orientation as shown by the pattern of the contours.

The low correlations of Figure 2D, the negative correlations of Figure 2E, and the large distances of Figure 2A-2C generally coincide. This may be interpreted as resulting from several possibilities. Where the structural features are more prominent and marked, larger changes usually take place between adjacent grid points and regularity is not general. As mentioned previously, the direction of and the amount of the changes are not necessarily the same for all the input maps. Conversely, in the Salina Basin where the units are nearly featureless, horizontal, and parallel, there is little difference between the adjacent grid points. Consequently, smaller distances and higher correlations are observed for most of the adjoining data points. Examples of changes that could affect the distances or correlations between neighboring points include (1) offset or migration of structural axes with depth which is usual in Kansas Plains–type folds; (2) truncation of beds by unconformities, such as the Pennsylvanian-Mississippian unconformity which causes differences in structural attitudes; and (3) variations of structural orientations of beds generated by changes in thickness of stratigraphic units because of differential compaction. Several hypothetical traverses across different types of structures were generated graphically to determine whether the postulated causes could produce correlations and distances that resemble those observed for the five structure contour maps. Simple graphical simulations were done for (a) differential compaction and (b) offsets of fold axes. In the simulations both positive and negative correlations were obtained, however the negative ones do not necessarily go with structural highs and positive ones with structural lows if the data are standardized. The simulations do suggest that one or more of these agents of change are at least plausible. Needless to say other explanations also can be visualized for these data and other samples where this technique might be used.

CONCLUSIONS

Our technique for the comparison of adjacent spatial data points taken from a series thematic maps is another aspect of spatial geostatistics; it combines the virtues of speed and simplicity and provides another way to examine the interrelationships of spatial data. The method discussed here can be included easily in any spatial analysis package for the investigator to visualize the lateral variation between data points on a group of maps. The method is programmed easily for a computer.

ACKNOWLEDGMENTS

We would like to thank Ute Herzfeld of Scripps Institute of Oceanography and Peter Sutterlin of Wichita State University for reading a preliminary version of the manuscript. Their comments were most helpful. We also would like to thank Sheryl Schrag and Linda Spurrier for typing the manuscript.

REFERENCES

Brower, J.C., and Merriam, D.F., 1990, Geological map analysis and comparison by several multivariate algorithms, *in* Statistical applications in the earth sciences: Geol. Survey of Canada Paper 89-9, p. 123-134.

Cole, V.B., 1962, Configuration of top Precambrian basement rocks in Kansas: Kansas Geol. Survey Oil and Gas Invest. No. 26, map.

Henderson, R.A., and Heron, M.L., 1977, A problematic method of paleobiogeographical analysis: Lethaia, v. 10, no. 1, p. 1-15.

Herzfeld, U.C., and Merriam, D.F., 1991, A map-comparison technique utilizing weighted input parameters, *in* Computer applications in resource estimations: Pergamon Press, Oxford, p. 43-52.

Journel, A., and Huijbregts, C., 1978, Mining geostatistics: Academic Press, New York, 600 p.

Merriam, D.F., 1960, Preliminary regional structural map on top of Mississippian rocks in Kansas: Kansas Geol. Survey Oil and Gas Invest. No. 22, map.

Merriam, D.F., 1963, The geologic history of Kansas: Kansas Geol. Survey Bull. 162, 317 p.

Merriam, D.F., and Smith, P., 1961, Preliminary regional structural contour map on the top of Arbuckle rocks (Cambrian-Ordovician) in Kansas: Kansas Geol. Survey Oil and Gas Invest. No. 25, map.

Merriam, D.F., and Jewett, D.G., 1988, Methods of thematic map comparisons, *in* Current trends in geomathematics: Plenum Press, New York, p. 9-18.

Merriam, D.F., and Sondergard, M.A., 1988, A reliability index for the pairwise comparison of thematic maps: Geologisches Jahrbuch, v. A104, p. 433-446.

Merriam, D.F., Winchell, R.L., and Atkinson, W.R., 1958, Preliminary regional structural contour map on the top of the Lansing Group (Pennsylvanian) in Kansas: Kansas Geol. Survey Oil and Gas Invest. No. 19, map.

Raup, D.M., and Crick, R.E., 1979, Measurement of faunal similarity in paleontology: Jour. Paleontology, v. 53, no. 5, p. 1213-1227.

Sneath, P.H.A., and Sokal, R.R., 1973, Numerical taxonomy: W.H. Freeman and Company, San Francisco, 573 p.

Watney, W.L., 1978, Structural contour map: base of Kansas City Group (Upper Pennsylvanian) - eastern Kansas: Kansas Geol. Survey Map M-10, map.

MAP INTEGRATION MODEL APPLIED IN SITE SELECTION

Elahe Tabesh
Syracuse University, Syracuse, New York, USA

ABSTRACT

Site selection is a location dependent decision-making task, and it is carried out traditionally by overlying various types of maps and attempting to assimilate number of factors into a composite picture. Geologists may be involved in various site selection jobs such as: exploration, environmental studies, construction, and drilling-site selection.

Such a decision can be one of the two types. The first is the unstructured decision—that is based on experience and an overall understanding of the problem. The second type of decision involves a certain amount of precise data and an algorithm to process it, so that a definite or structured decision can be reached. Analysis of the elements of a geologic problem identifies the spectrum the problem belongs. Structured (well-defined) versus unstructured (ill-defined) decision making for site selection will be discussed. To keep the location dependency of variables and taking them into account during the process of decision-making, the variables treated in layers or maps. A map integration model will be introduced that carries the decision-making task and results in the most suitable location.

Use of Microcomputers in Geology, Edited by D.F. Merriam
and H. Kürzl, Plenum Press, New York, 1992

INTRODUCTION

What is a decision? A decision is the result of a process of selecting a specific course of action from a number of possibilities. The consequences of the action should be preformulated and appraised. The relevant existing site selection models in the literature are based on determining a mathematical solution to the problem, and have made use of statistical decision theory. Many researchers have approached this concept in context of mineral exploration. Two major methods are used more than the others in applications: "Bayesian" and "non-Bayesian" (classical). Statisticians of both schools agree on the fundamental rules. The difference is that the Bayesian approach provides a formal mechanism for taking the preferences into account instead of leaving it to the intuition of decision maker. However, without formalization, a decision made under uncertainty has remained essentially arbitrary. The formalization of utilities and weights lead to decisions that are arbitrary but in some ways more objective. In most applied decision problems, both the preferences of a responsible decision maker and the judgment on the weights to be attached to the various possible states of nature are based on substantial objective evidence. The quantification of personal preferences and judgment enables the decision maker to arrive at a decision which involves these objective evidences.

STRUCTURED (WELL-DEFINED) DECISION MAKING

Decision Parameter

To frame a decision problem the following parameters must be identified for each individual situation.

1) Space of feasible actions: A={a}.
   The decision maker selects a single act a from domain A, which includes a set of potential acts. For site selection, a will be the subarea the most suitable for certain purpose.
2) State space: S={s}.
   There must be at least two different "states of the world": the consequence of adopting act a depends on some state of the world. The selection or response of the decision maker is to accept one or another hypothesis of the possible state of the world. In site selection problem, this variable will be represented by subareas with higher potential.

3) Family of resulting events: E={*e*}.
   The evidence or information that the decision maker uses is denoted by *e* from a family E. For a site selection, it is the consequence of certain decisions in terms of the involved factors.
4) Utility function: u(., ., .,).
   The decision maker assigns a utility function u(*e*,*a*,*s*) to perform a particular *e*, taking a particular action *a*, and then determining that a particular *s* obtains. The function u considers the cost associated with certain decision in the different subareas under the prevailing conditions.

In addition to the given parameters Bayesian decision making has an extra parameter.

5) Probability assessment: P{ ., . | *e*}
   For every *e* in E the decision maker directly or indirectly assigns a probability to show the likelihood of evidence *e* in the set of E.

Rendu (1976) reviewed this method in the context of mineral exploration. A complete example of framing a site selection problem in a structured decision-making model was given and approached by the Bayesian method. He concluded that

Statistical analysis appears to be a logical tool for optimization of decision in mineral exploration. However, geologists might find it difficult to quantify their opinion, decision makers might consider that a utility function gives an over-simplified representation of their preferences. Some aspect of the statistical decision theory, such as the use of a decision tree and rigorous structurization of the decision process will be more easily accepted and can be used as a starting point for introduction of the theory in an exploration company.

The problems that Rendu (1976, p. 443) mentioned are general and they apply not only to Bayesian decision-making models but also to all classical decision-making methods.

Definition

Given the set of feasible actions, A, the set of relevant states, S, the set of resulting events, E, and a (rational) utility function, u that orders the space of events with respect to their desirability the optimal decision under uncertainty is the choice of the action leading to the event with the highest utility; such a decision can be described by the quadruple {A,S,E,u}. It should be noted that A, S, and E are sets, and u is a function that induces an order on *e*. The components of the basic model under certainty are taken to be *crisp* sets of function. Crisp indicates dichotomous (selection of one and only one decision of X, which is considered to be "best").

The set of action, set A can be defined precisely if the possible state (or states) and the utility function u are precise. Vagueness enters the picture only when considering decisions under risk, which occurs if one of the mentioned elements is not precise or uncertainty concerns the occurrence of a state or the event itself.

Geologists constantly work with nature and its complexity; therefore, all of the factors that they have to consider in the decision-making model inherit the complexities to some extent. Most geologic data are imprecise and incomplete; so that geologists must make decisions in an uncertain world based on inferential reasoning. Objective decision-making models are not efficient enough to accommodate these complexities, or to consider the approximate and incompleteness nature of geologic data.

This problem can be resolved by imitating the human decision-making process. There are both major differences and similarities between human and formal reasoning and decision making. Both procedures are based on probabilistic reasoning. With questions about the probability that event A originates from process B, the degree to which A is representative of or resembles B is evaluated, although ignoring prior probabilities, the effect of sample size and the principle of regression to the mean are ignored. Another heuristic method used by a decision maker assesses the probability of an event based on the ease with which instances or occurrences can be reminded. In both the logical and statistical domains, it seems that human reasoning is dependent on context, so that different operations or inferential rules are required in different context.

UNSTRUCTURED (ILL-DEFINED) DECISION MAKING

Generally, making certain decisions relies completely on the procedures of decision making rather than on the objective of thought. These are subjective; there exist in the mind of a person thinking rather than to the objective problems. Any subjective evaluation or decision is the result of a conscious classification. This type of decision is dependent on context and is considered to be information processing. Usually it is difficult to define the sets A, S, and E, and even the utility function is considered to be developed within the decision process. Decisions are treated dependent of context, and the analysis must include the human being as the decision maker. This is termed unstructured decision making. Precision is no longer assumed, and ambiguity and vagueness may be modeled only verbally, which usually does not permit the use of mathematical methods for analysis and computation. Most human decisions are considered to be subjective. Classical decision-making models are not adequate to model such problems.

Fuzzy set theory provides a mathematical framework in which vague phenomena can be studied precisely and rigorously. It serves as a modeling language for situations in which fuzzy relations, criteria, and phenomena exist. Because geologic information is descriptive and usually described by natural language, fuzzy set theory can help formulate complexities, incompleteness, and uncertainties associated with geologic problems.

THE CONCEPT OF FUZZY SET THEORY

A crisp set is defined normally as a collection of elements or objects $x \in X$ which can be finite and countable. Each element can either belong or not belong to sets A, A † X. The statement that "x belongs to A" is true; in the former situation but not in the latter. Such a set can be described in different ways, one can list the elements that belong to the set. analytically for instance, by stating conditions for membership ($A=\{x \in X \mid x \leq 5\}$) similarly one can define the member elements by using the characteristic function 1/A, in which 1/A(x)=1 indicates membership of x to A and 1/A(x)=0 nonmembership. In a fuzzy set the characteristic function allows the degree of membership for the elements of a given set to differ between 1 and 0.

Example 1

A geologist is searching for the best location in an area to set up a drilling rig. A suitable site may be on a potentially host rock for a certain type of mineralization. Assume that the area consists of eight different rock types with different potential likelihood with respect to being the host rock. Let X={1,2,3,....,8} be the label of these rock types. Then, the fuzzy set "favorable subarea based on the rock types" will be described as

$$\tilde{A}= \{(1, .5), (2, .2), (3, .8), (4, 1), (5, .7), (6, .3), (7, .5), (8, .9)\}$$

(favorability ranges from 0 to 1 with 1 being most favorable). Obviously, the judgment of which rock type is most favorable for different types of mineralization is based on geologist knowledge, experience, and observations.

A fuzzy set is a generalization of classical sets and the membership function is a generalization of characteristic functions. Because we refer to a crisp set X, some elements of a fuzzy set may have zero degree of membership. It may be appropriate to consider those elements in the universe that have a nonzero degree of membership in a fuzzy set.

Definition

The crisp set of elements that belongs to the fuzzy set A, at least to the degree a, is termed the "a -level set."

$$A_a = \{x e X \mid m_A(x) \geq a\}$$

$A'_a = \{x e X \mid m_A(x) \geq a\}$ is referred to as a "strong a-level set" or "strong a-cut." Again, with reference to the example 1 and listing the possible a-level sets:

$$\tilde{A}_{.2} = \{1,2,3,4,5,6,7,8,\}$$

$$\tilde{A}_{.5} = \{1,3,4,5,7,8,\}$$

Note that the rock types 2 and 6, with respective degrees of membership .2 and .3, are present in $\tilde{A}_{.2}$ but not in $\tilde{A}_{.5}$.

$$\tilde{A}_{.8} = \{3,4,8\}$$

$$\tilde{A}_1 = \{4\}$$

Example 2

Let A be the fuzzy set from example 1 and B be the fuzzy set of the rock types with favorable fracture systems as favorable features for mineralization:

$$\tilde{B} = \{(1, .2), (3, .5), (4, .3), (6, .4), (7, .7)\}$$

The intersection $\tilde{C} = \tilde{A}$ h $\tilde{B}$ is then

$$\tilde{C} = \{(1, .2), (3, .5), (4, .3), (6, .3), (7, .5)\}$$

The union $\tilde{D}=\tilde{A} \cup \tilde{B}$ is

$$\tilde{D}=\{ (1, .5), (2, .2), (3, .8), (4, 1), (5, .7), (6, .4), (7, .7), (8, .9) \}$$

The complement of $\complement\tilde{B}$, which is interpreted as "area with insufficient fracture density," is

$$\complement\tilde{B}= \{ (1, .8), (2, 1), (3, .5), (4, .7), (5, 1), (6, .6), (7, .3), (8,1) \}$$

Decision Making Under Fuzzy Conditions

In a fuzzy decision model (Bellman and Zadeh, 1970) the objective functions and their constraints are characterized by their membership functions. Because the purpose is to satisfy (optimize) the objective functions and constraints, a decision in a fuzzy environment is defined by analogy to a nonfuzzy environment as the selection of activities that simultaneously satisfy objective function(s) and constraints. According to the given definition and assuming that the constraints do not interact, the logical "and" corresponds to the intersection. The decision in a fuzzy environment therefore can be viewed as the intersection of fuzzy constraints and fuzzy objective function(s).

Example 3

Objective function x should be substantially larger than 10 and characterized by the membership function

$$m_{\tilde{G}}(x)= \begin{cases} 0 & x<10 \\ (1+(x-10)^{-2})^{-1} & x>10 \end{cases}$$

Constraint x should be in the vicinity of 11, and characterized by the membership function

$$m_{\tilde{C}}(x)= (1+(x-11)^{4})^{-1}$$

The membership function $m_{\tilde{D}}(x)$ of the decision is then

$$m_{\tilde{D}}(x_{max})= m_{\tilde{G}}(x) \;\; m_{\tilde{C}}(x)$$

$$= \begin{cases} \min\{1+(x+(x-10)^{-2})^{-1}, (1+(x-11)^{4})^{-1}\} & \text{for } x>10 \\ 0 & \text{for } x<10. \end{cases}$$

The fuzzy decision is characterized by its membership function for all xeX:

$$m_{\tilde{D}}(X)= \min\{m_{\tilde{G}}(x), m_{\tilde{C}}(x)\}$$

If the decision maker intends to make a crisp decision, it seems appropriate to suggest the state with the highest degree of membership in the fuzzy set "decision." This is termed the maximizing decision x_{max} with

$$m_{\tilde{D}}(x_{max})= \max_x \min \{ m_{\tilde{G}}(x), m_{\tilde{C}}(x)\}$$

In the previous situation the min-operator was used based on the following argument: in the classical (crisp) selection model of decisions the verbal linkage between constraints and goals usually is "and." The intersection of fuzzy sets, however, has so far been modeled or defined by a min-operator. The question arises whether the association "and" - "logical and" - "intersection" - "min-operator" lead to an appropriate model for decisions.

Bellman and Zadeh (1970) indicated that their interpretation of a decision is more general in several ways. One possible generalizations involves logical operators, as illustrated in the following example.

Example 4

A geologist intends to drill a hole in an area that has been divided into a number of subareas. An evaluation based on geological and geophysical factors for one of the subareas shows two different favorability factors or degrees of membership in the fuzzy sets "favorable geological condition (G)" and "favorable geophysical condition (P)", in comparison with the other sets. Assume that the favorability was evaluated as follows:

$$m_{\tilde{G}}=.8 \qquad \text{and} \qquad m_{\tilde{P}}=.5$$

If the decision by the geologist corresponds to the degree of membership of the fuzzy set "best location for drilling" it would be conceivable that the geologist's best decision m_D could be determined by

$$m_{\tilde{D}}= \max(m_{\tilde{G}} , m_{\tilde{P}}) = \max (.8, .5) = .8$$

The generalization makes it easy to include the selection model as well as the evaluation model in the notion of a fuzzy decision. In a sense, this has already been done in example 4.

Decision Making and Georeferenced Data

Decision making, similar to other operations in the context of georeference data, is location-dependent, and somehow this dependency must be taken into account. Here it can be argued that X and Y coordinates can be treated as two individual constraints; this can be valid for a single-criterion decision making. When comparing different solutions in terms of desirability, judging the suitability, and determining the "optimal" solutions multiple phenomena generally should be used. This has led to the subject of Multi-Criteria Decision-Making (MCDM) in the framework of numerous evaluation schemes that have been suggested, one of which concentrates on decision making with several phenomena in Multi-Attribute Decision-Making (MADM). In examples 1 and 2 selecting the most satisfactory location for a drilling rig is a decision that is being affected by several variables. The coordinate data differ from the rest of the attributes; and should not be treated in same way as the rest of the data. Yet, at the same time, all the attributes are depend on location, and they have to be treated according to geographic locations. MCDM and MADM models are not adequate for multiattribute decisions that are applicable to spatially distributed data. The inherited uncertainty and subjectiveness also should be modeled somehow.

To consider location dependency the data should be treated as layers. Each attribute is represented by one layer. To summarize: a proper decision-making model for site selection should be able to address the following issues:

1) Uncertainties associated with nature.
2) Vague and incomplete information.
3) Descriptive information, and the difficulties of converting them into numerical format, also subjectiveness.
4) Inhomogeneities in the nature of data.
5) Spatial-dependency of data.

To address these issues a map integration model is developed based on adapting subjective decision making. The model works the best in association with an expert system or it can be built in most GIS (Geographic Information System) packages running on microcomputers. The model will be described in relation to an example.

APPLICATION

The model is tested by applying to a data set from Big Hill silver-zinc deposit, Pembroke, Maine. The area has a long history of exploration, a detailed geologic map is available, geophysical and geochemical surveys were carried out. Resulting maps show the distribution of several elements in the soil. The goal is to consider the locality of metallic anomalies, and the other controlling factors that affect the economy of exploration, and design an "optimal" pattern for exploratory drilling. The map integration model is applied to this data set in order to select the best or the optimize drilling sites.

Geology

Figure 1 shows the geological map of the area. The area is underlain by Silurian volcanic rocks typical of the Machias-Eastport volcanic rocks of Northern Appalachian region. The basalts are reported to be part of the Leighton Formation of Silurian-Early Devonian age (Gates and

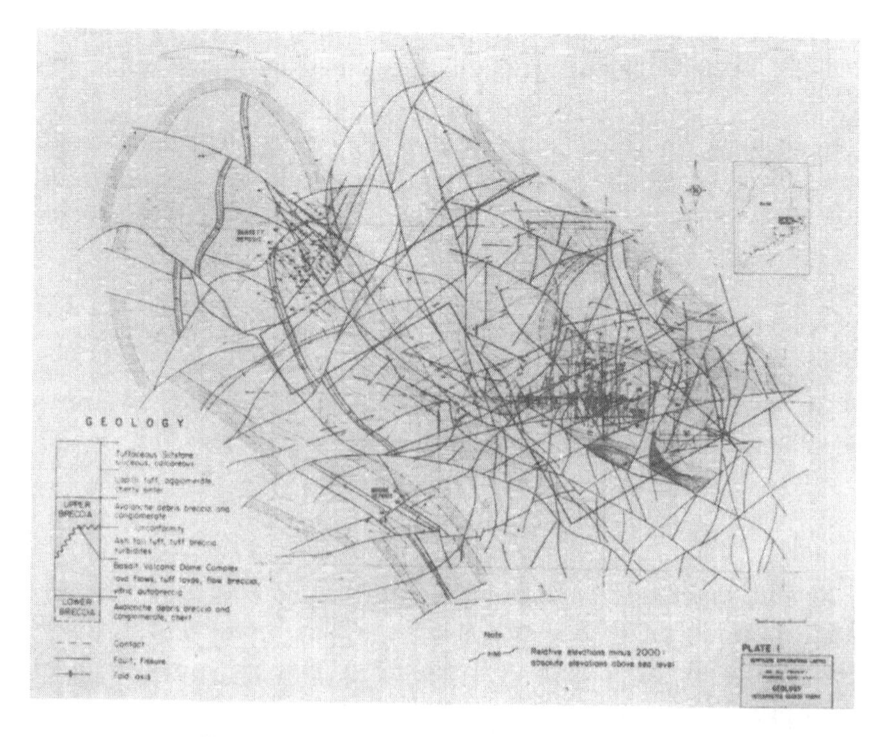

Figure 1. Geologic map of Big Hill area.

Moench 1981). They are composed of basalt flows, coarse basaltic agglomerate, tuff breccia, and ash fall deposits. The dip of stratigraphic sequence averages 30 degree to the northeast. The sequence is repeated at the surface several times because of shallow folding and normal faulting. A dense well-ordered fissure pattern is located at the Big Hill area and seems to partially control the mineralization.

The mineral enriched flow traveled upward along channels, and was deposited at certain elevations, by filling the fissures (Schaff 1982). In brittle rocks, such as basalt flows, quartz-carbonate-sulfide mineralization occurs continuously for hundreds of feet as stockwork and shatter breccias. Where fissure feeders transacted porous stratigraphic horizons, such as avalanche breccia and conglomerates, tuff breccia, and flow breccia. The quartz-carbonate-sulfide mineralization saturates the pores of the primary rock and open spaces which are structure related.

The low-sulfide primary ore consists of stringers, disseminations and coarse clotty aggregates of low-iron sphalerite, galena, pyrite, chalcopyrite, and minor tetrahedrite. Native silver was reported present as inclusions in all of the sulfides, including pyrite, but mainly in galena and sphalerite. Quartz and carbonate are the gangue minerals. The pertinent geological factors include a map of distribution of host rocks: basalt, breccia and ashfall, and a map of distribution of fracture and faults.

Map Digitization and Integration

A map layer is a two-dimensional data set showing the distribution of a single variable or attribute. Individual layers of digitized map were produced for the pertinent factors. In the host rock map geologist decided to differentiate between various type of host rock by assigning various favorability factor with respect to their suitability of being host rock. The following set of factor represents the suitability of host rocks: basalt (1), upper breccia (0.7), lower breccia (0.5), and ash fall (0.6). Of course these factors are subjective and are based on geologist judgment, but the purpose is to model subjectivity in decision making. The area can be divided into grided subareas, and based on dominant rock type a favorability factor will be assign by the system to each cell. The factors are treated as continuous variables and are shown by contour lines. Figure 2 is the result of an evaluation of the area based on host-rock distributions with respect to higher chance of hitting the orebody.

The map layer showing the distribution of faults and fractures is treated differently. The distribution of fractures is a favorable factor in

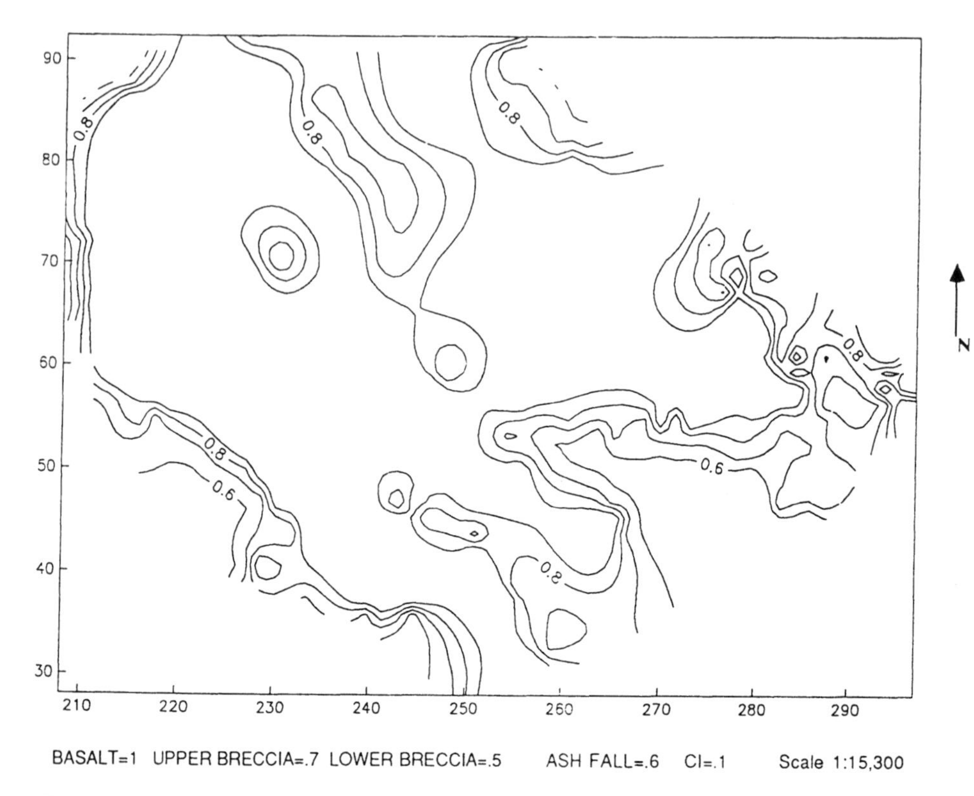

Figure 2. Contour map showing rock favorability with respect to mineralization in Big Hill area.

porphyry and stockwork deposits. Therefore, the area with the highest fracture densities is the most favorable area for mineralization. The distribution of fractures also is treated as a fuzzy set, and the degree of membership is represented by favorability factors. Based on the assumption that the cell with the highest fracture density is the most favorable area for mineralization, the rest of the cells were compared with that one and ranked. The system overlays a grid mesh on the top of the map, and then searches for the cell with the highest fracture densities and assigns 1 to the most favorable subarea for mineralization. The search continues and each grid cell is evaluated depending on its relation to the cell with the highest density; and favorability factors ranging from 0 to 1 are assigned (Fig. 3). Figure 4 is the contour map showing the distribution of densities in Big Hill. Finally, the maps representing the favorable area based on the presence of host rock and fracture densities will be integrated by applying the logical operator. The grids that are laid over each map layer are differed and it is a

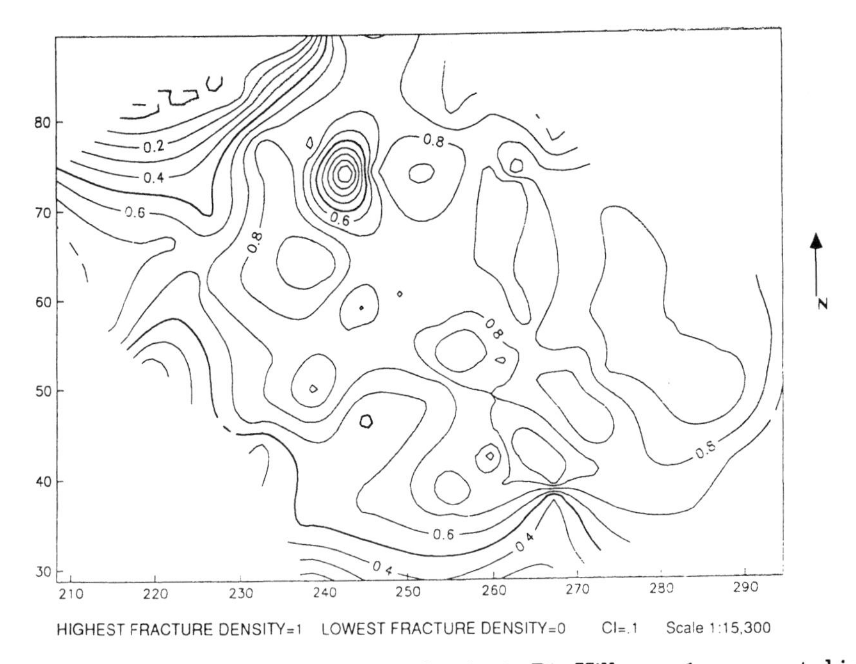

Figure 3. Contour map showing fracture density in Big Hill area; 1 represents highest density.

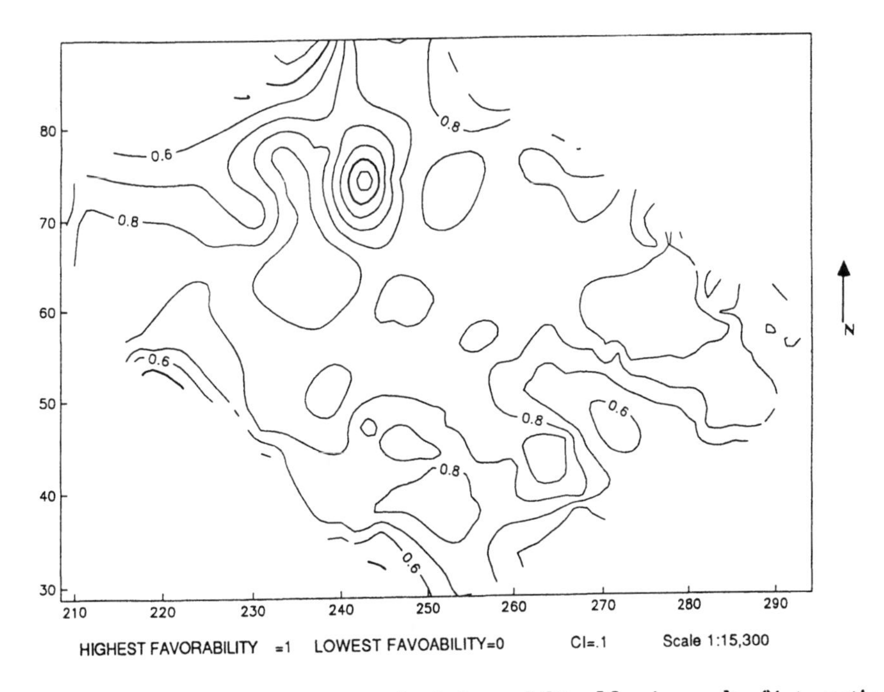

Figure 4. Contour map showing geologic favorability. Map is result of integrating maps in Figures 2 and 3.

function of the size of the feature on each layer. For the purpose of integration, the finer sized grids have to be converted to the largest size, and the values assigned to the smaller cells will be added and then divided by the number of grid cells (Fig. 5). If the larger grid covers some fraction of the smaller ones, the area of each fraction acts as a weight for the value assigned to it (Fig. 6).

An ideal situation for porphyry deposits is the presence of host rocks and a dense fracture system. To integrate these two map layers representing two complementary variables, the logical operator AND

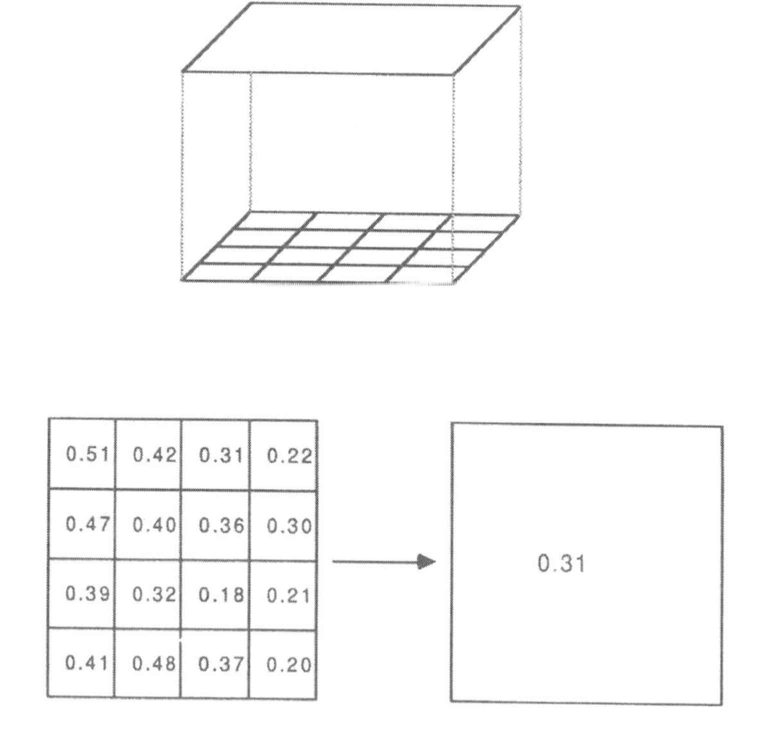

0.51+0.42+0.31+0.22+0.47+0.40+0.36+0.30+0.39+0.32+0.18+0.21
0.41+0.48+0.37+0.20/16= 0.31

Figure 5. Converting smaller cell values to larger one. Larger cell lies over an even number of smaller cells. Because smaller cells occupy equal proportion of area, arithmetic mean will represent value for large cell.

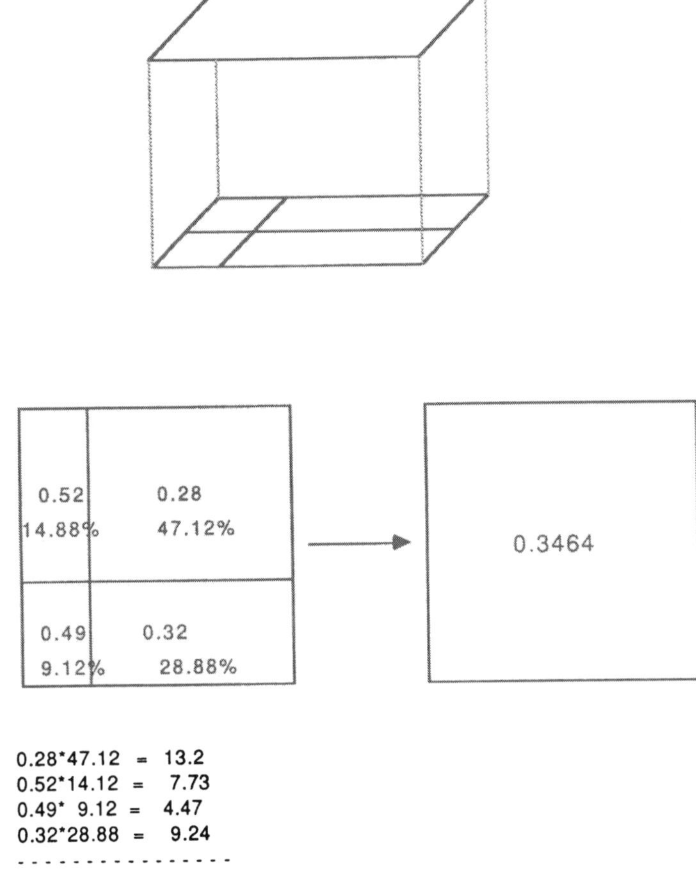

Figure 6. Converting smaller cell values to larger one. Large cell does not overlay even number of smaller cells. Value of each smaller cell weighs proportionate to area that it occupies.

was used. This is an emulation of the selecting process of an area for a specific goal (drilling) with respect to the geology, that an expert geologist follows.

Processing Exploration Data

Exploration data refers to the data gathered through geophysical and geochemical surveys. In the Big Hill area geochemical analyses of the soil were carried out for four elements: lead, zinc, silver, and copper. The distributions are shown in Figures 7-10. Spontaneous Polarization (SP) surveys were carried out over the area, the negative anomalies indicate the location of silver concentrations (Fig. 11).

The user should establish a second map file for the exploration data, digitize the maps, and store them. The system evaluates each individual map layer based on instruction provided for various data sets. For example, the user should make it clear that on the SP map the negative anomalies are favorable and that they indicate silver concentrations. The system divides the area into subareas by overlaying an appropriately sized grid mesh, evaluates each subarea, and assigns favorability factors to them. These evaluated map will be stored in a secondary file. Figures 12-16 show the evaluated maps for each data layer. It is obvious in these maps that the areas showing higher favorability factors coincide with the anomalies in the original maps. The map integration process follows the same procedures that were followed for integrating the host rocks and fracture density maps. The map shown in Figure 17 indicates the area with the highest favorability with respect to geochemical data. Figure 18 shows the result of integrating Figure 17 with the SP-evaluated map (Fig. 16). It should be remembered that SP anomalies indicate the silver concentration; as a result the silver factor in Figure 18 already is weighted. If there is an interest in specific elements because of higher market price or higher demand the data can be evaluated with a weighting factor that expresses the level of interest.

Processing Engineering Factors

Because of the nature and subjectivity of some of the engineering factors, they could not be taken into account so far, but they are important in exploration capital cost. The approach should be analytical and starts by the indication of various attributes. In the Big Hill area these factors are recognized as the physiography of the area and accessibility to the road. For these two factors the topographic map of the area was digitized

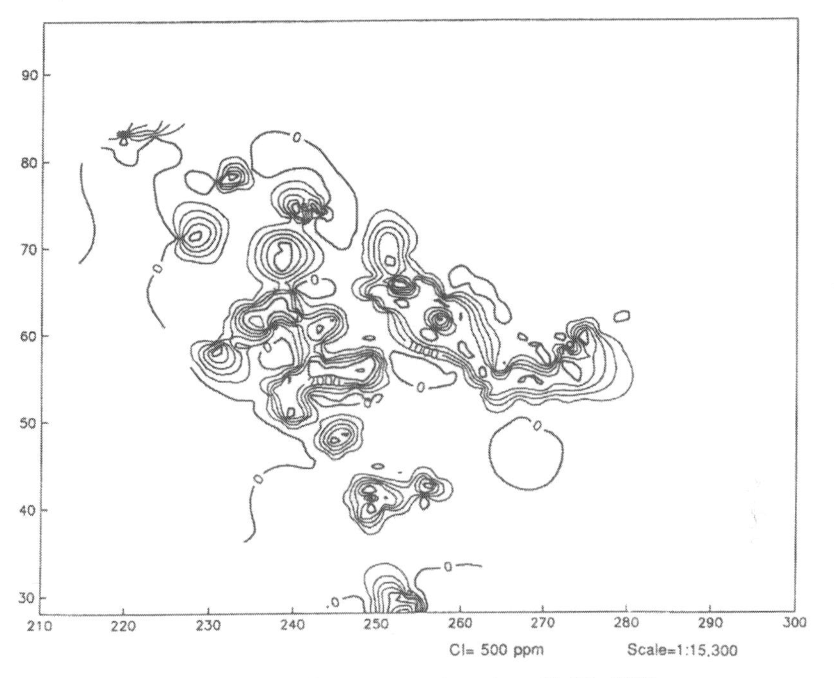

Figure 7. Lead concentrations in soil, Big Hill area.

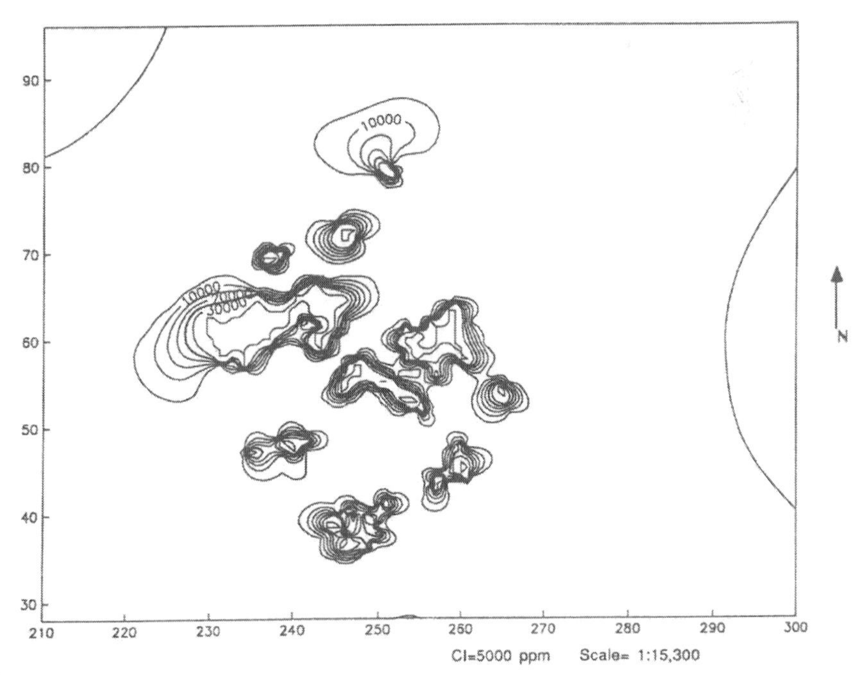

Figure 8. Zinc concentrations in soil, Big Hill area.

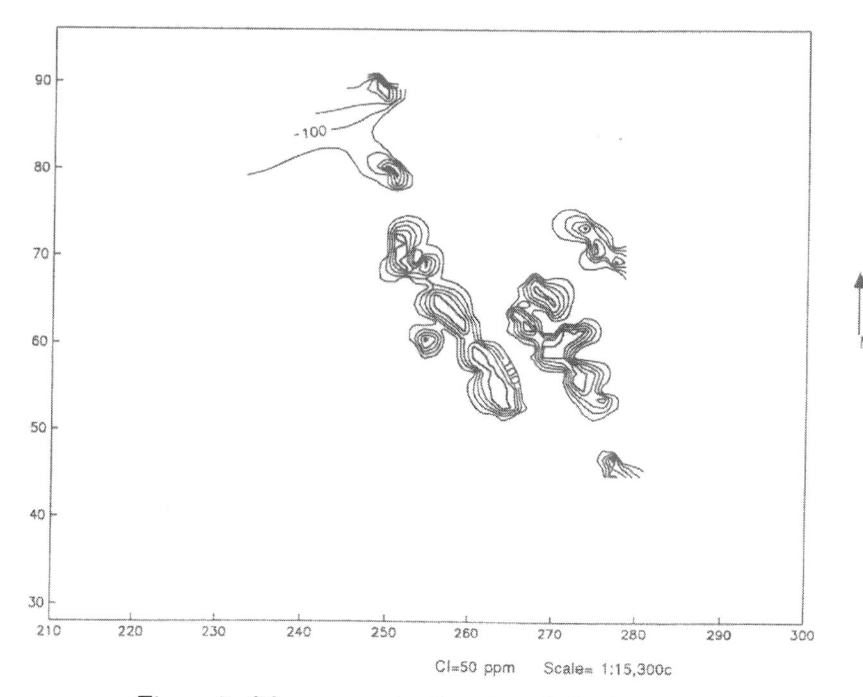

Figure 9. Silver concentrations in soil, Big Hill area.

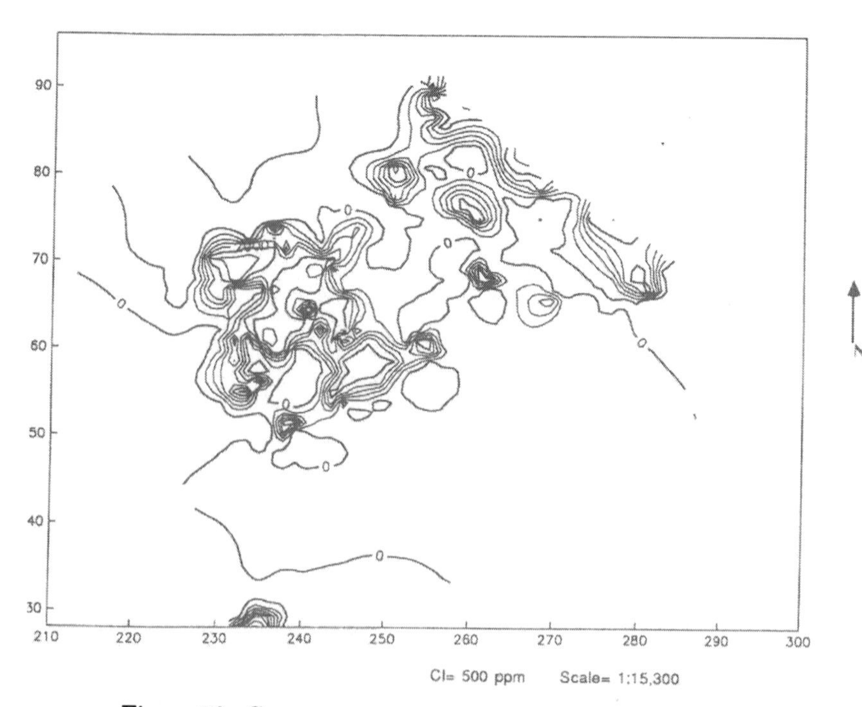

Figure 10. Copper concentrations in soil, Big Hill area.

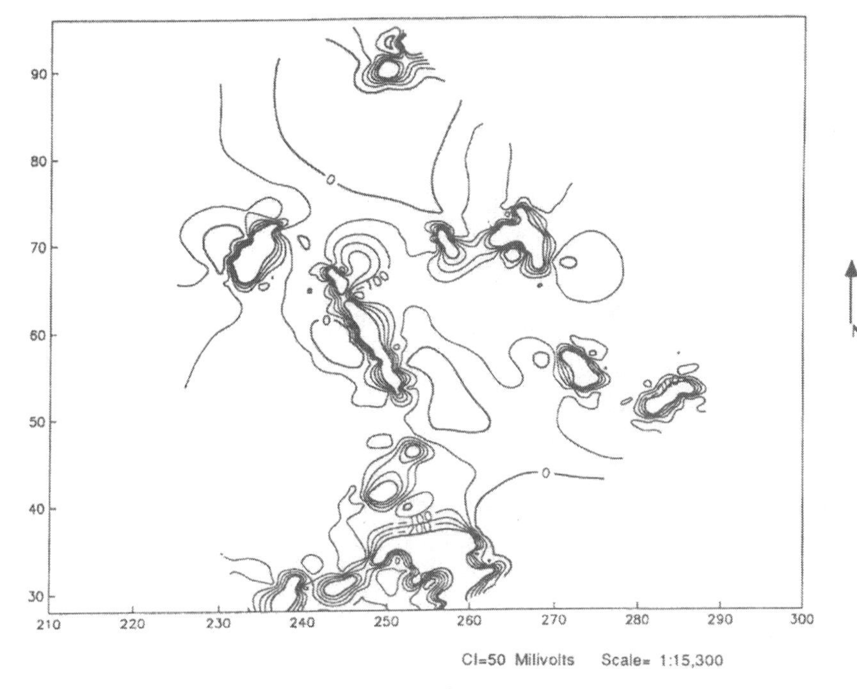

Figure 11. Results from spontaneous polarization surveys, Big Hill area.

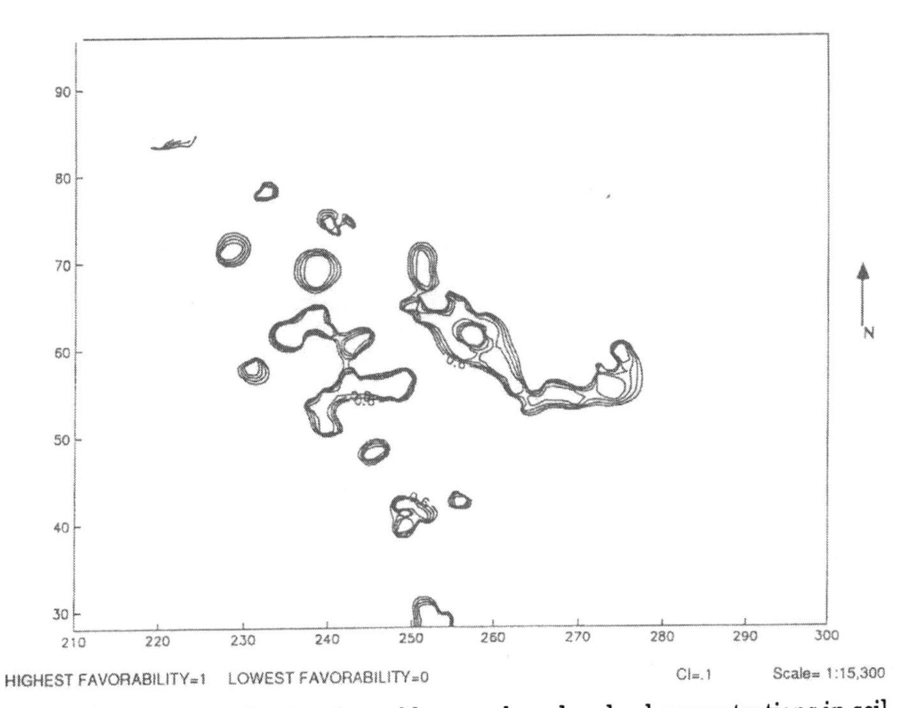

Figure 12. Contour map showing favorable areas based on lead concentrations in soil, Big Hill area.

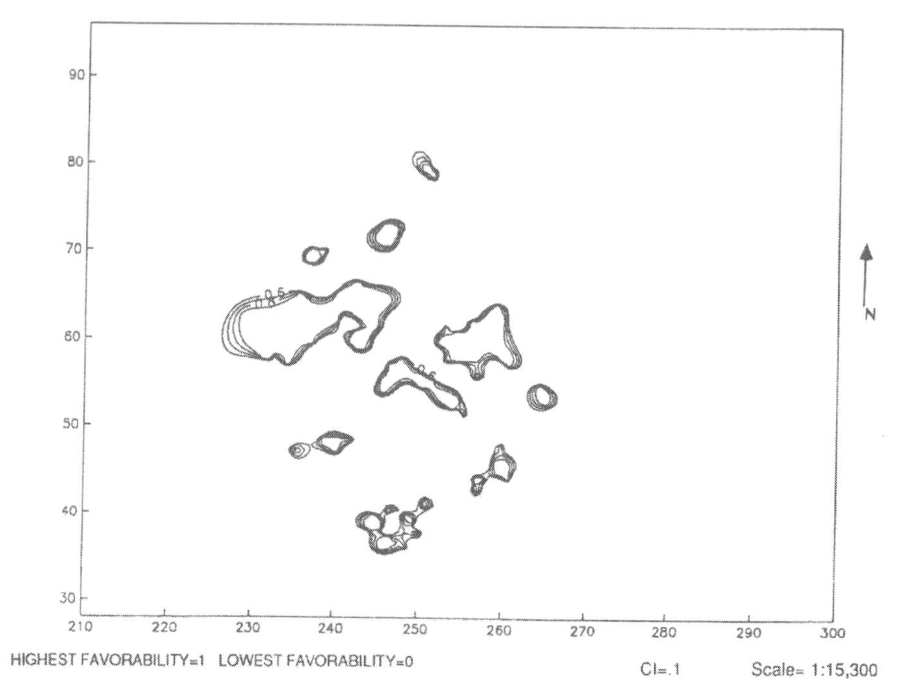

Figure 13. Contour map showing favorable areas based on zinc concentrations in soil, Big Hill area.

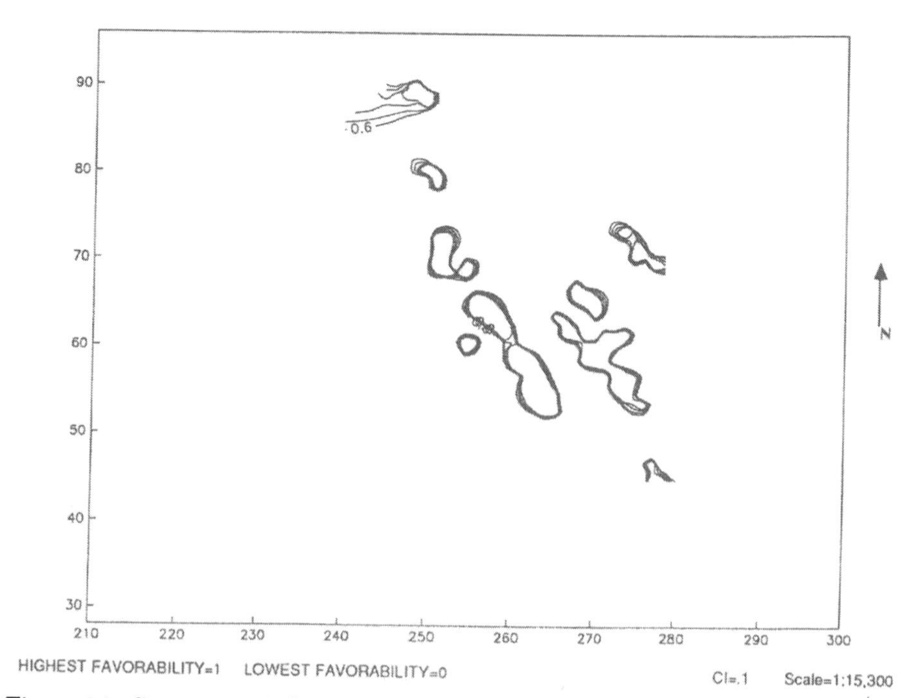

Figure 14. Contour map showing favorable areas based on silver concentrations in soil, Big Hill area.

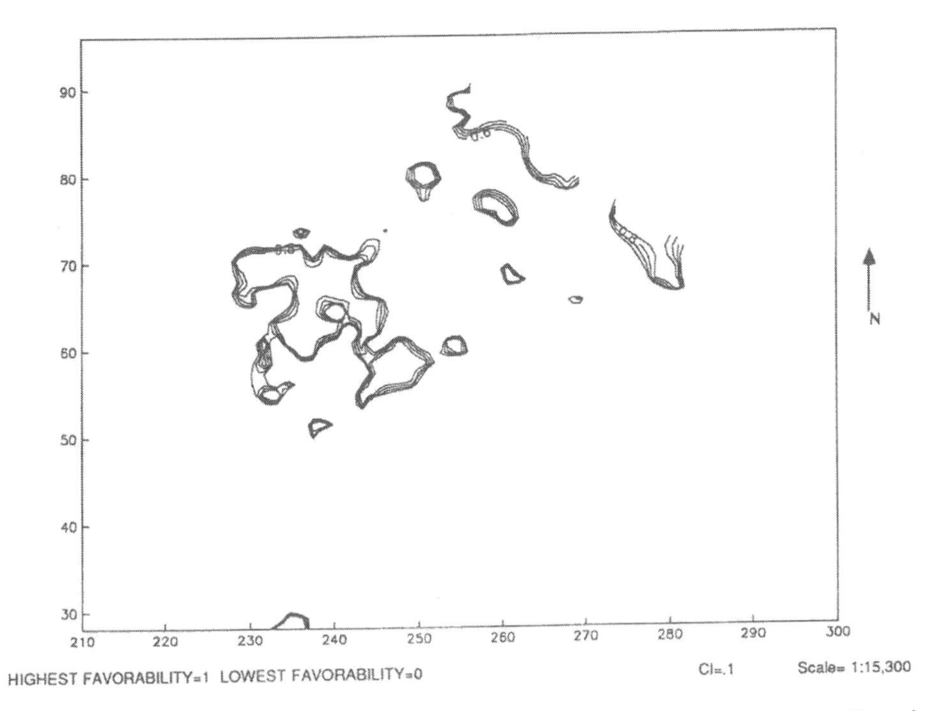

Figure 15. Contour map showing favorable areas based on copper concentrations in soil, Big Hill area.

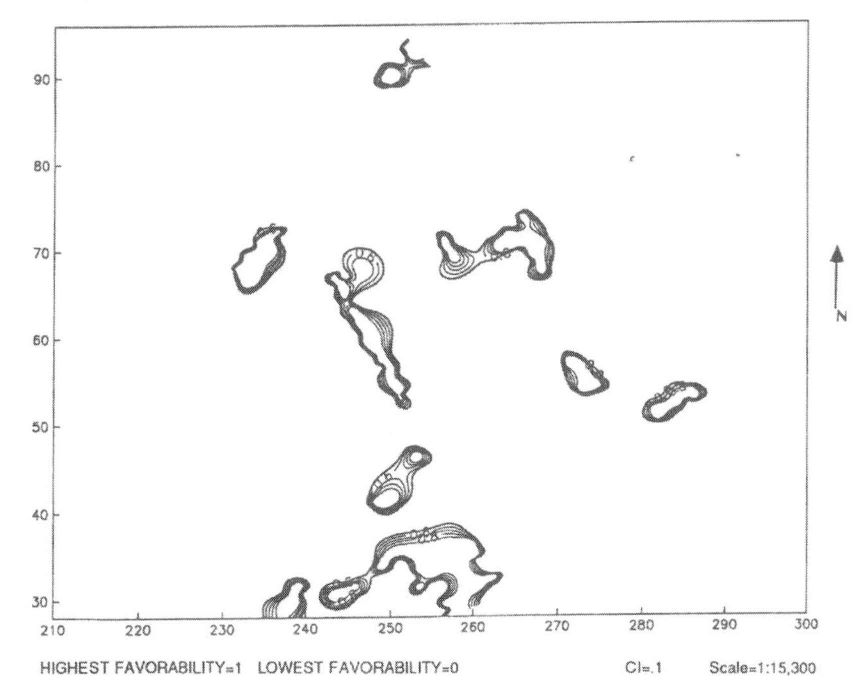

Figure 16. Contour map showing favorable areas based on SP surveys, Big Hill area.

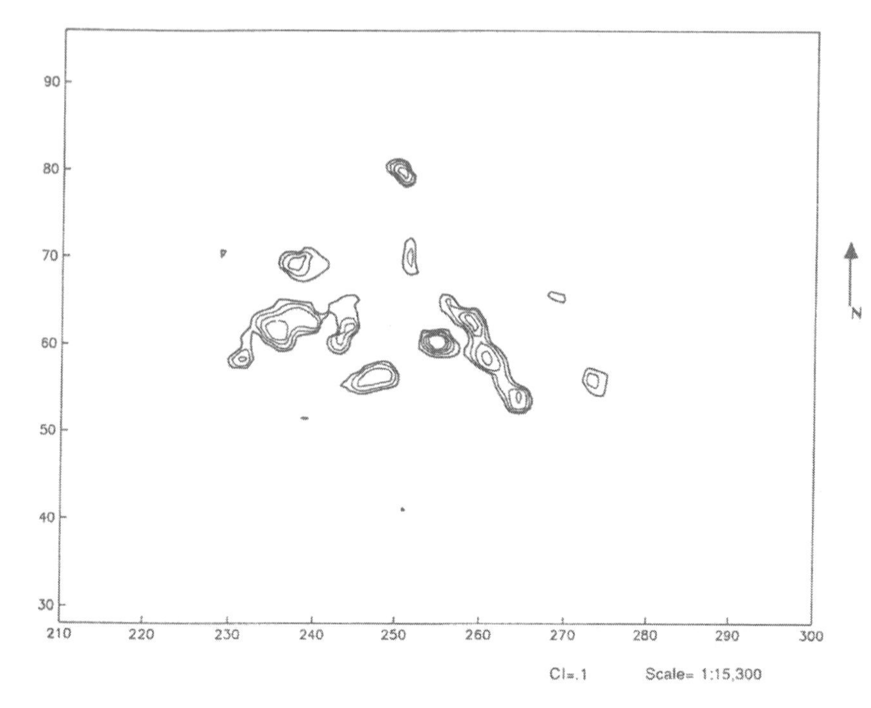

Figure 17. Contour map showing favorable areas based on concentrations of lead, zinc, silver, and copper in Big Hill area. Map is result of logical integration of Figures 12-15.

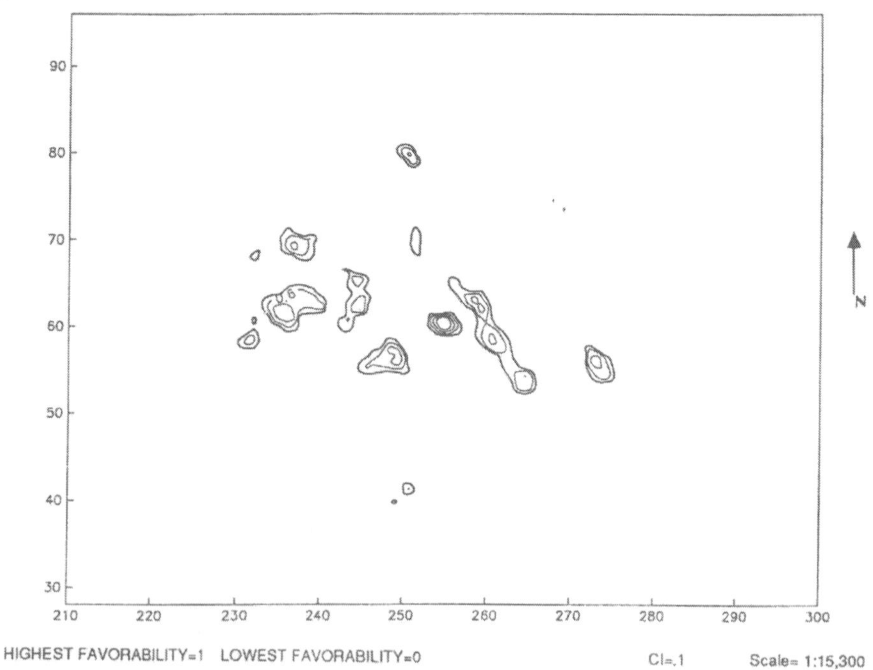

Figure 18. Contour map showing result of integrating Figure 17 with SP surveys (Fig. 16).

from Figure 1 and reproduced by computer (Fig. 19). A second map layer was produced by digitizing the roads accordingly to their ranking. The system subdivided the area by overlying a grid mesh on the map. For example it calculates the distance between the center of each cell to the road. Cells on topography map will be evaluated based on their suitability for setting up the rig. The factors that were considered in evaluating the area include slope and accessibility to the area. Figure 20 shows the favorability based on these two factors.

Final Integration and Results

The maps representing favorability based on three groups of factors (of geological, exploration, and engineering) are integrated by applying the OR operator. The action of this operator is similar to comparing these map layers and selecting the area with the maximum favorability factors. The integration of maps within each group of factors involved the application of the AND operator. The action of this operator is similar to comparing various map layers and selecting the grid cell with the minimum favorability factors. The combination of AND and OR operators result in the "max-min" (maximum of the minimum value), which is logically the best or optimum selection under the prevailing conditions. This is the essence of unstructured decision making that allows ill-defined constraints to be taken into account.

By comparing the maps indicating favorable areas with respect to geology (Fig. 4) with those showing exploration data (Fig. 18), it can be seen that the favorable exploration area coincides with part of the geologically favorable area (at the middle of the map). Figure 21 was produced by laying Figure 18 over Figure 1 and it shows the anomalous area located next to a fault. Along the same fault there is another anomalous area toward the north. Because the exploration data consist of soil analyses, these anomalies might be an indication of the fault being a channel-way to silver-enriched flow and of the concentration of the elements along it therefore, the groundwater flow should be studied in the area and be considered as an important factor. Two other anomalies occur in the area that already has been drilled. Based on these results, it seems that drilling had priority in the indicated areas rather than around Mains Zone and the Barrett deposits. Another favorable area falls on the south of the Barrett deposit and a small one north of the Moore deposits.

Figure 22 is the result of integrating Figure 18 and Figure 20 (favorable area based on engineering factors). This figure confirms the

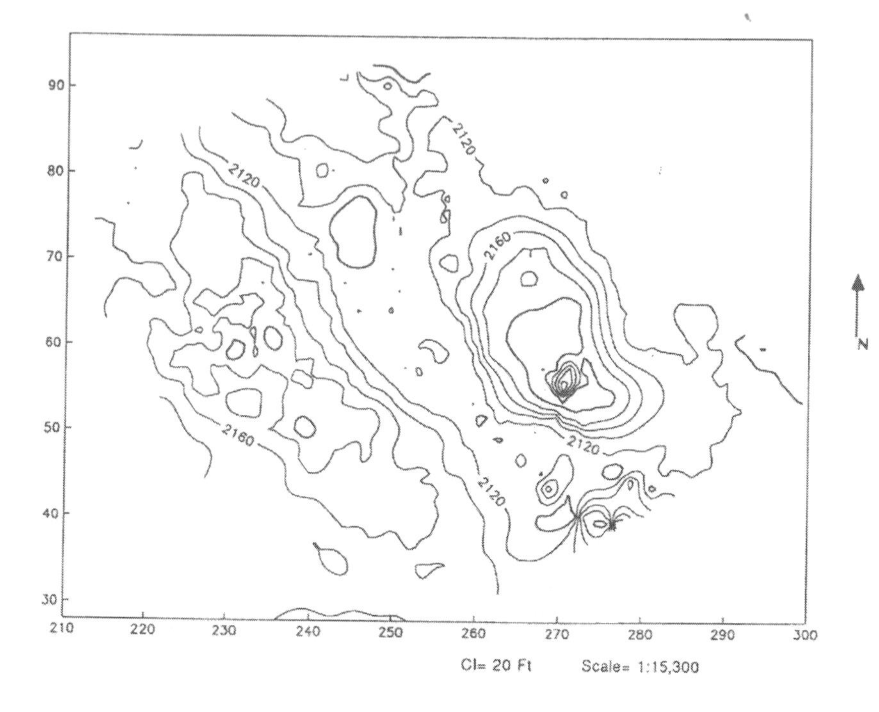

Figure 19. Computer reproduction of topographic map of Big Hill area.

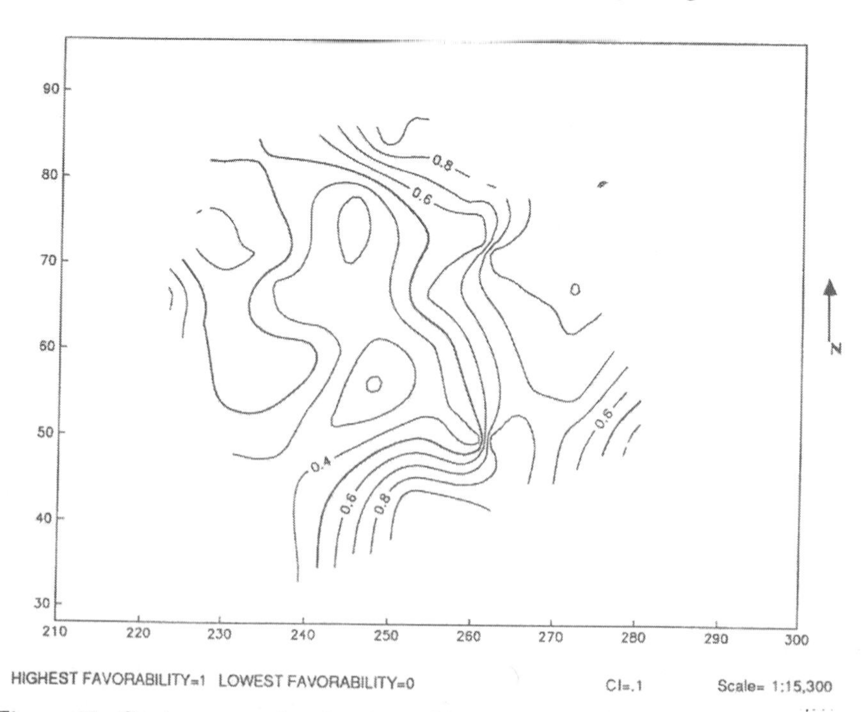

Figure 20. Contour map showing favorable area based on engineering factors (favorablility based on topography and accessibility).

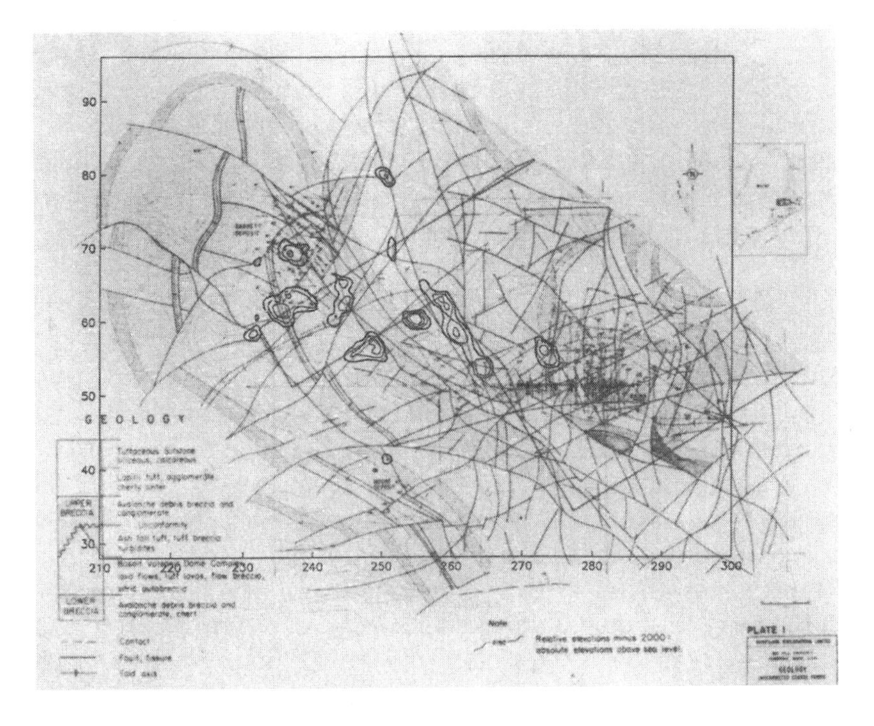

Figure 21. Figure 18 overlaid on Figure 1.

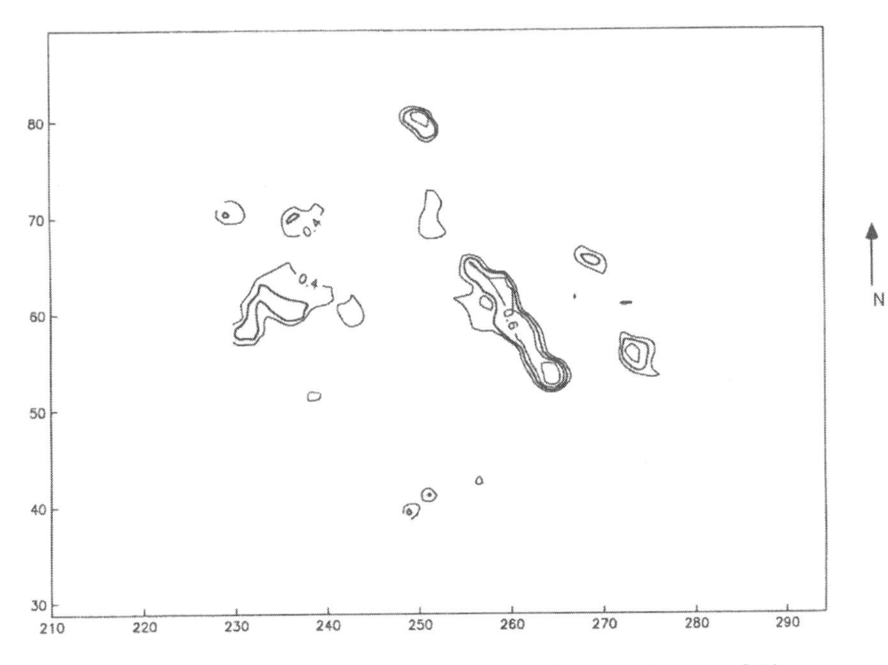

Figure 22. Result of integrating maps from Figures 4, 18, and 20.

areas indicated in Figure 18 as highly favorable areas. In addition to those areas, there are a few new areas on the eastern side of the maps. Because these anomalies first occur on this map they cannot be reliable and have to be examined in detail before being recommended for drilling. Visually comparing Figure 22 with Figure 2 (the geologically favorable area), it can be detected that the appearance of these anomalies is the result of the favorable geological circumstances of the area.

CONCLUSIONS AND RECOMMENDATIONS

The introduced map integration/site selection model is able to consider location dependent variables from various sources, and assists geologists to make a site selection decision. Because this model is based on the use of logical operator, the technique can be added simply to any GIS package. The system is based on subjective decision making which allows considering features such as: vague and incomplete data sets, uncertainties associated with geological variables, and inhomogeneous data. Because this model emulates human decision making, the best application for it, is to extend that by designing an interface between the GIS and an expert shell. Then the concepts for evaluation and assigning the scale factors can be constructed in form of rules in the knowledge base of expert shell, and the whole evaluation process can be done automatically.

The application of the system requires a rigorous amount of digitization. If a file of scanned map data can be converted into a digital file that is compatible to the system, then the digitization will not be necessary.

REFERENCES

Bellman, R.E., and Zadeh, L.A., 1970, Decision-making in fuzzy environment: Management Science, v.17 B, p.141-164

Gates, O., and Moench R.H., 1981, Bimodal Silurian and Lower Devonian volcanic rock assemblages in the Machias-Eastport area, Maine: U.S. Geol. Survey Prof. Paper 1184, 32 p.

Rendue, J.M., 1976, Bayesian decision theory applied to mineral exploration and mine evaluation, *in* Advanced geostatistics in the mining industry: D. Reidal Publ. Co., Dordrecht, p. 435-445.

Schaff, R.P., 1982, Status report on Big Hill Silurian-Zinc deposit, Pembroke, Maine, USA: internal report for Scintilore Exploration Limited, 15p.

ANALYSIS OF SPACE AND TIME DEPENDENT DATA ON A PC USING A DATA ANALYSIS SYSTEM (DAS): A CASE STUDY

Rudolf Dutter and Gerald Karnel
Institute for Statistics and Probability Theory, Vienna, Austria

ABSTRACT

The program system DAS (Data Analysis System) is designed for the interactive graphical analysis of spatial data on the PC. It has no special features for the treatment of spatial time series but it is so flexible that it can be used for the analysis of such data. Some properties of the program and a case study on pollution data in Austria are presented to illustrate its capabilities.

THE PC PROGRAM DAS: DATA ANALYSIS SYSTEM

DAS is an easy to use PC program for exploratory, numerical, and graphical statistical data analysis. It is both (optionally) command and menu driven. The graphics modules include histogram, cdf-plot, one-dimensional scattergram, density trace, boxplot, xy-diagram, ternary diagram, and draftman's display. Statistical tables including mean, standard deviation, robust mean (Huber and Hampel), robust scale, median, percentiles, hinges (quartiles), interquartile range, medmed, letter values, and tabulation of upper and lower outliers by result and sample number can be calculated and printed. Sorted data tables, multielement-outlier tables and raw data tables can be prepared as well.

Use of Microcomputers in Geology, Edited by D.F. Merriam
and H. Kürzl, Plenum Press, New York, 1992

Additionally, DAS offers two special features:

- the use and interactive definition of data subsets in almost all graphics, and
- the possibility to use regionalized variables and to combine the geographical distribution of samples with their statistical analysis.

These two features make DAS an ideal instrument for all scientists working with regionalized data (e.g. environmentalists, geochemists, biologists, toxicologists) who want to map the geographical distribution of samples and results in addition to a modern statistical analysis.

Data subsets can be predefined or interactively created in diagrams and maps or by using simple mathematical operations. Up to 100 data subsets or subsubsets can be saved in the original data file under a unique name. The geographical distribution of any subset can be studied, subsets can be compared in most graphics and a full statistical analysis can be carried out for any one, or all, data subset(s).

For the special geographical displays, digitized topographical maps can be used as background information. These topographical background maps can be converted from AUTOCAD (tm)\footnote{Any trademarks mentioned here are trademarks of the respective companies.}, ARC/INFO (tm)[1], or simple ASCII datafiles.

Further support during data analysis is offered by a data transformation module (log, logit, box-cox, multiplication, ladder of power, range, etc.), a case selection module and a data definition facility. The latter can be used to define new variables via mathematical operations (e.g. $FeO=Fe \times 1.286$ or $A={Na}_{2} O+K_{2}O)$.

The data set for DAS can be entered either directly into DAS via screen forms, or taken from dBase III/IV (tm) or Lotus 1-2-3 (tm). Of course, DAS also can create directly dBase III/IV or Lotus 1-2-3 files. Additionally, several ASCII data file formats can be used for data input/output.

All graphical outputs of DAS can be prepared in publication ready quality. Almost every parameter influencing the graphical presentation on the screen can be modified.

Each of the diagrams, maps, etc. can be arranged on the screen in an arbitrary fashion. A worksheet of any predefined size (usually A4 or

---

[1] Any trademarks mentioned here are trademarks of the respective companies.

A3 but, also up to 3m by 3m) is divided interactively into windows which will contain further on the diagrams, maps, explanations, or text. Any once-designed worksheet layout can be saved into special "sheet"-files using a unique name.

A text facility can be used to write, move, modify, and save annotations for the layout. More than 20 different text fonts are available. A set of more than 100 symbols for mapping such as north arrows, scale bars, symbols for castles, churches, towns, etc. is included. Each symbol can be positioned anywhere on the screen in any size. It also is easy to create new symbols and add them to the set.

A command tracing can be involved where all used commands (typed or generated via the menu) will be written on a file. This file can be used to repeat parts of or the complete session of your analysis.

A "SNAPSHOT" command can be issued at any time. Whenever the user approves the graphical layout on the screen, "snapshot" will write the present layout to a meta file. This meta file can be displayed at another time, plotted, printed by a laser printer, or converted to be read by AUTOCAD (tm) for further graphical processing. It may be edited in a simple way to make final corrections. Interfaces to popular desktop publishing programs such as Ventura Publisher (tm) or Page Maker (tm) exist as well as interfaces to word-processing programs such as WORD 5.0 or TeX (see also Dutter and others, 1990).

THE CASE

The data we are using have been collected in an industrial area of Austria where stack sulphur and dust have been collected and the pH-value has been measured at 34 spots around a chimney. Each observation corresponds to a period of 28 days, that is one year consists of 13 units. The original data have been stored on separate files for each year and spot so we had to reorganize them by gathering all 170 files (34 spots in 5 years) in one large file to be able to begin with the analysis. The original data file consisted of 6 variables, three of them were SO_2, SO_3, and sulphur per day. After some calculations we noticed that these variables had been constructed by multiplying the measured sulphur with a constant. Thus, from the statistical point of view these variables are of no further interest and we will focus on sulphur, dust, and the pH-value.

THE ANALYSIS

Univariate Statistics

In this section we will compute some basic statistics such as the median, mean, variance, and some robust measures of our three variables and we will plot histograms, density and cumulative density functions, and boxplots of each variable.

Sulphur - Table 1 is taken directly from the output of DAS. Sulphur is measured in mg. First we can see the number of missing values being 49 out of 2048. Different location estimates are provided as well as estimates of variation. Skewness and kurtosis are large, that is the bulk of the data is narrow and the larger values produce a heavy tail.

Table 1. Univariate analysis of sulphur

```
+======+
! S mg !
+======+

Number of values used   : 1999
Number of missing values : 49
Name of subset(s)       : NONE

Boxplot Statistics                Location Estimates
Median :       1.172              Mean :      1.6645
Hinges :   0.581    2.217           Huber :     1.39103
Whiskers:   0.005    4.667           Hampel:     1.52019
Fences :  -1.873    4.671
Extremes:   0.005   11.088          Coeff. of Variation: 0.969096

Variance    : 2.60195
St. Dev.    : 1.61306
Hinge-Spread:   1.636  (Normal Consistent: 1.21275)
MedMed      :   0.706  (Normal Consistent:  1.0467)
Robust Variance (Huber):  1.2883

Value        Value/S.E.
Skewness: 2.03958       37.2282
Kurtosis: 5.24043       47.8264

Minimum Standard Score: -1.02879    Maximum Standard Score:  5.84201
```

In Figure 1 three pictures show the shape of the data. The first one is a histogram combined with a scatterplot which is spread randomly in the y-axis, the second one shows a density function computed via kernel estimation combined with a boxplot and the third one shows a cumulative distribution function of the data in the logarithmic scale.

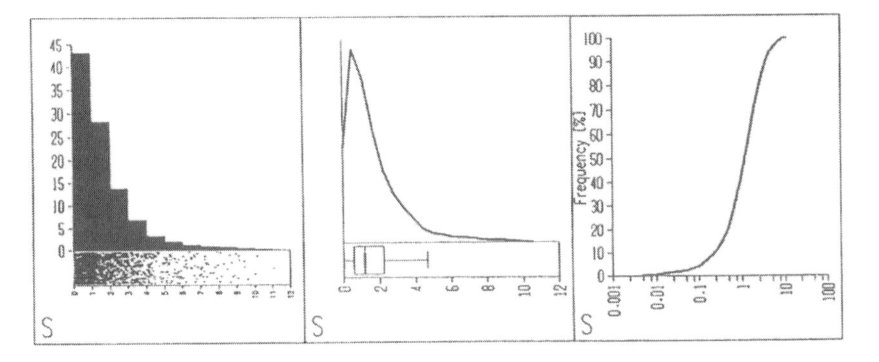

Figure 1. Univariate plots of sulphur.

Dust - Dust is measured in mg similarly to sulphur; 106 values are missing and again the skewness and the kurtosis are large (see Table 2).

The data are presented Figure 2 in the same way as Figure 1.

pH-value - The pH-value is missing in nearly one-third of the samples, skewness and kurtosis show a different behavior than the other two variables investigated. Here the bulk of the data is spread out and the tails are rather short (see Table 3).

In Figure 3 the data are presented in the same way as Figure 1 and Figure 2, only the scale of the cumulative density function is not a logarithmic one.

Time Series

In this section we want to look at our data in dependence of time. Without using any time-series smoothing procedure, Figure 4 and Figure 5 clearly show seasonal fluctuations the result of domestic fuel in winter.

On the other hand, dust and pH-value do not indicate any seasonal fluctuations (see Fig. 6 and Fig. 7).

Table 2. Univariate analysis of dust

```
+==========+
! Dust mg !
+==========+

Number of values used   : 1942
Number of missing values : 106
Name of subset(s)       : NONE

Boxplot Statistics              Location Estimates
Median :        119             Mean :     196.992
Hinges :    57.1    227.2         Huber :     142.665
Whiskers:    0.1    480.8         Hampel:      172.255
Fences : -198.05   482.35
Extremes:    0.1    3157          Coeff. of Variation: 1.30194

Variance   : 65777.5
St. Dev.   : 256.471
Hinge-Spread:  170.1  (Normal Consistent: 126.093)
MedMed     :     73  (Normal Consistent: 108.228)
Robust Variance (Huber): 14363.6

Value        Value/S.E.
Skewness: 4.15411      74.7356
Kurtosis: 28.8522     259.536

Minimum Standard Score: -0.767696   Maximum Standard Score: 11.5413
```

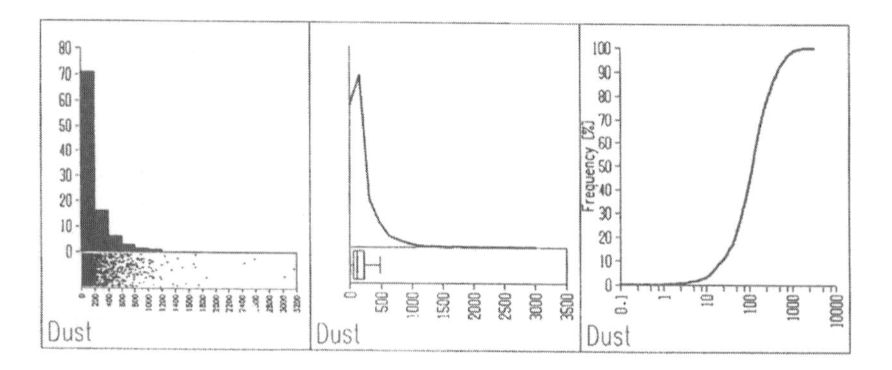

Figure 2. Univariate plots of dust.

Table 3. Univariate analysis of pH-value

```
+==========+
! pH-Value !
+==========+

Number of values used    : 1397
Number of missing values : 651
Name of subset(s)        : NONE

Boxplot Statistics              Location Estimates
Median :      6.26              Mean :     6.05548
Hinges :   5.01    6.95          Huber :     6.06441
Whiskers:   3.15    9.57          Hampel:      no convergence
Fences :   2.1     9.86
Extremes:   3.15    9.57          Coeff. of Variation:  0.1985

Variance    : 1.44484
St. Dev.    : 1.20201
Hinge-Spread:    1.94  (Normal Consistent:  1.4381)
MedMed      :   0.88  (Normal Consistent: 1.30467)
Robust Variance (Huber):  1.76559

Value        Value/S.E.
Skewness: -0.197919      -3.02002
Kurtosis: -0.704905      -5.37803

Minimum Standard Score: -2.41717    Maximum Standard Score:  2.92386
```

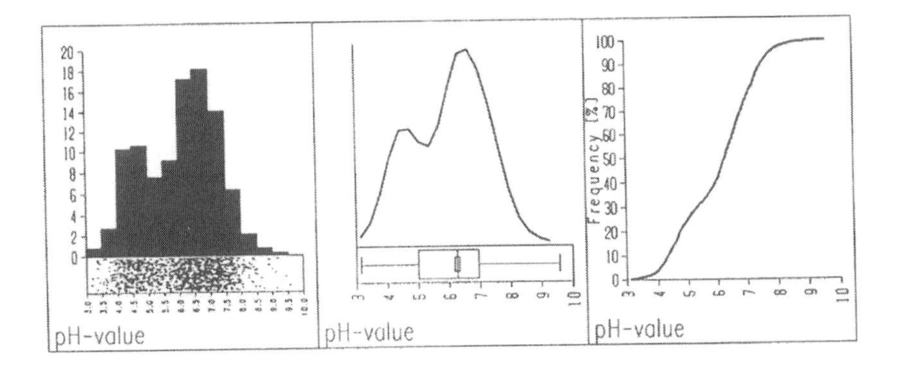

Figure 3. Univariate plots of pH-value.

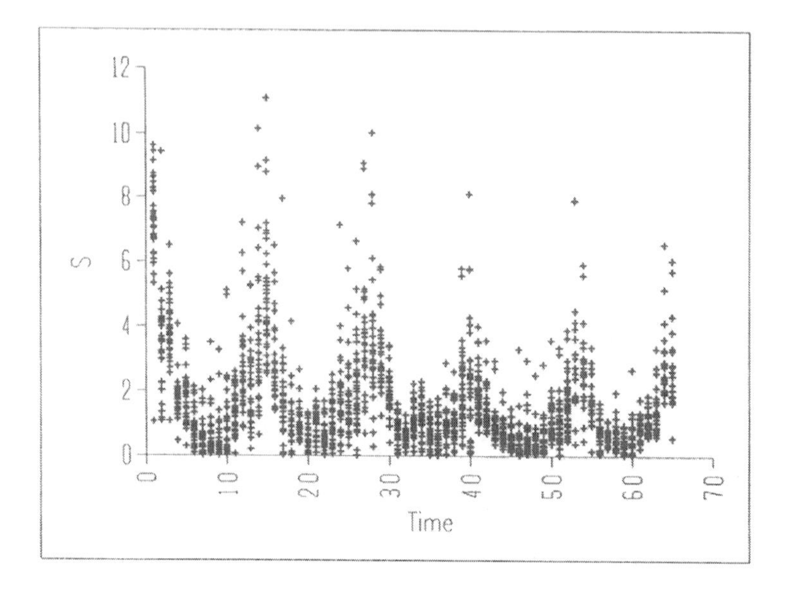

Figure 4. Sulphur through time.

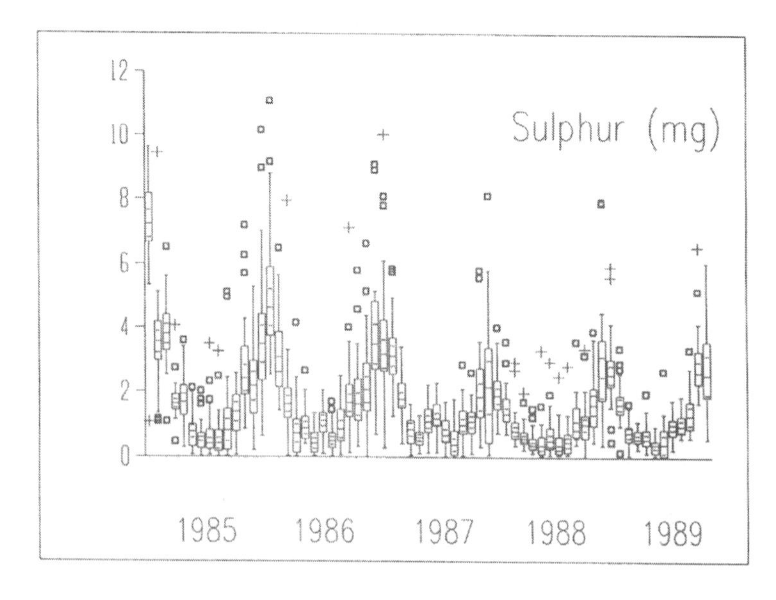

Figure 5. Boxplot of sulphur through time.

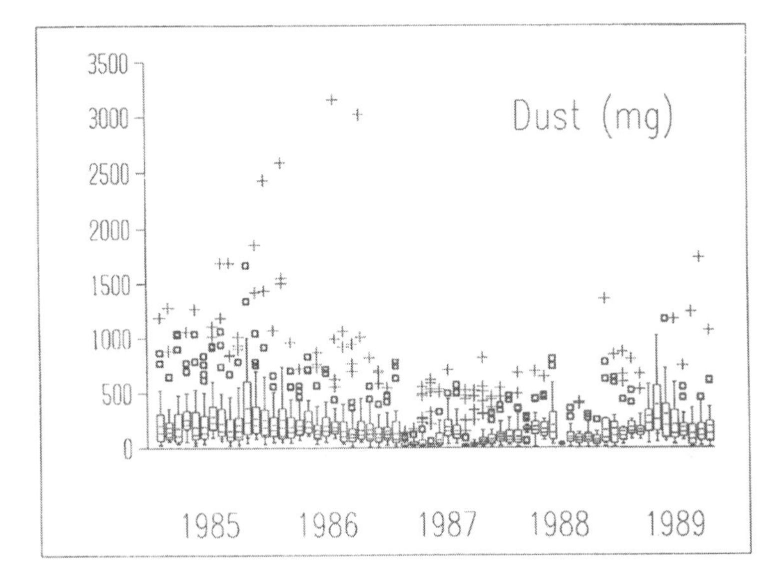

Figure 6. Boxplot of dust through time.

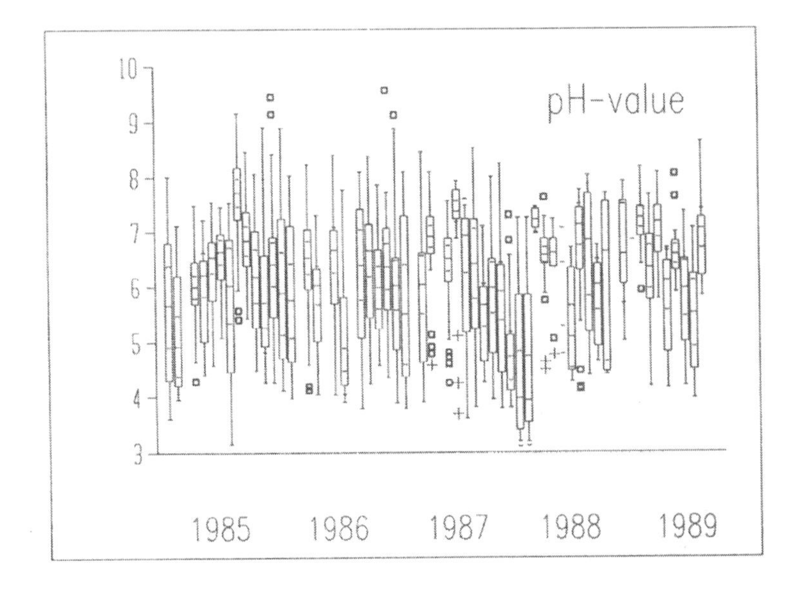

Figure 7. Boxplot of pH-value through time.

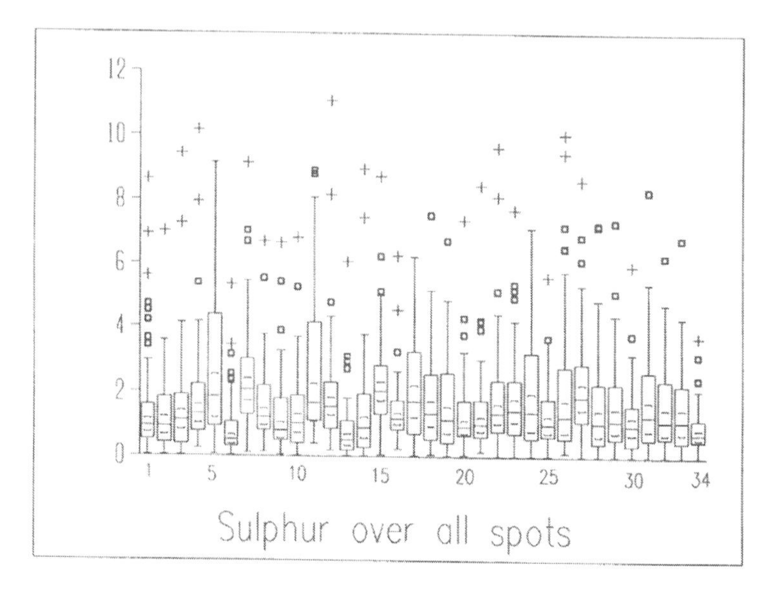

Figure 8. Sulphur over all spots.

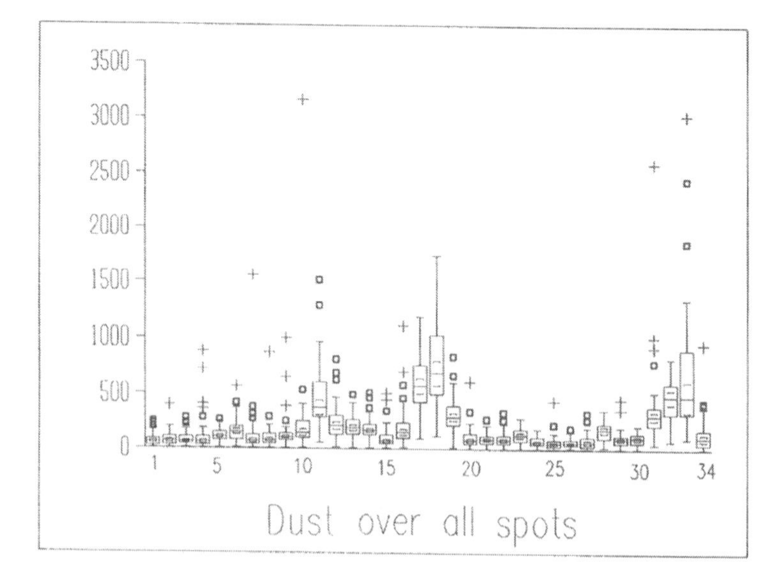

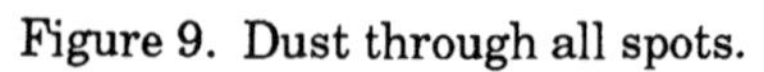

Figure 9. Dust through all spots.

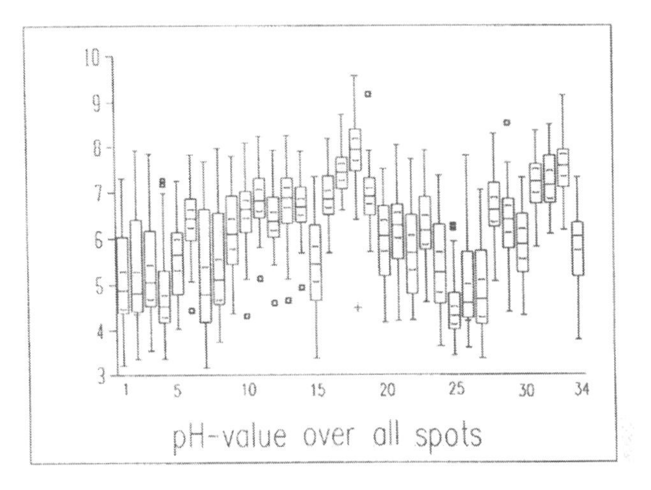

Figure 10. pH-values through all spots.

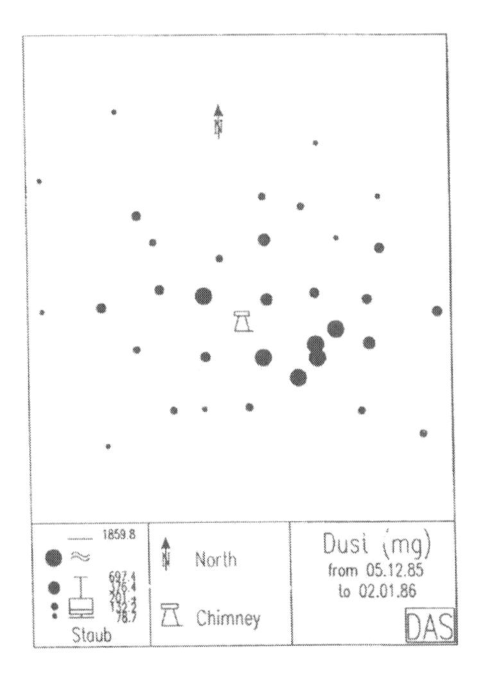

Figure 11. Dust from 5.12.85 to 2.1.86.

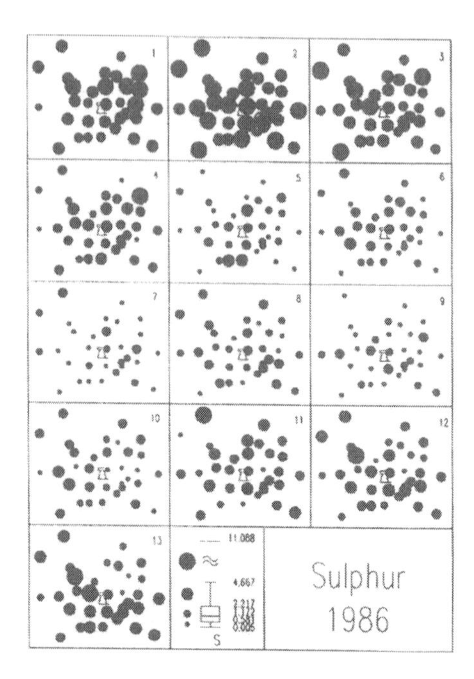

Figure 12. Sulphur in 1986.

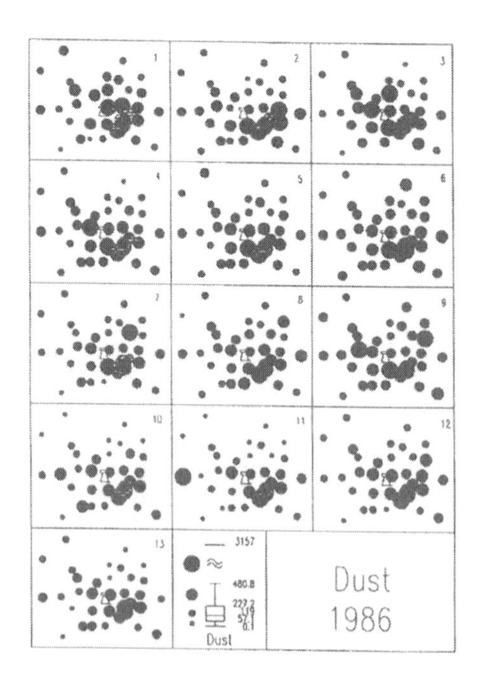

Figure 13. Dust in 1986.

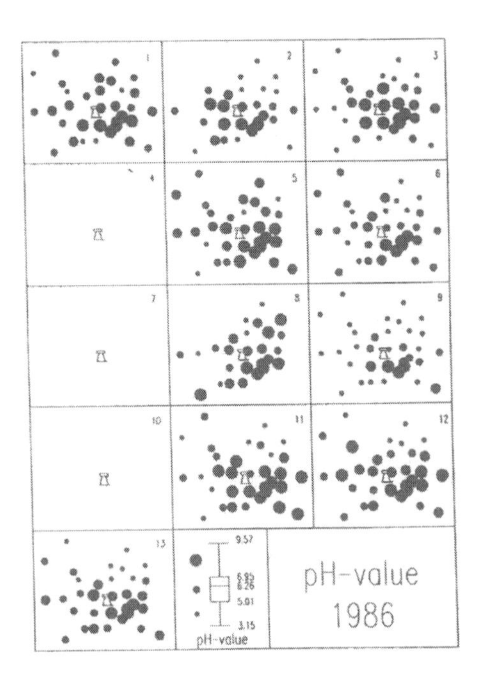

Figure 14. pH-value in 1986.

Visualizing Space Dependent Data

A simple way to visualize spatial data is to plot a variable against all spots, that is their indices without considering the exact geographical location, similarly to the time series. In our example we have 65 values at each spot (5 years \'a 13 units). With sulphur one easily can detect some large values indicated by '+' (see Fig. 8). These observations can be identified easily interactively by using the mouse or the cursor. Some spots (such as 6, 13, 34) seem to be the ones with the smallest amount of sulphur measured, whereas other spots (such as 5 and 11) show high 75% percentiles, and 12 shows a high maximum.

With dust, only some spots (11, 17, 18, 32, 33) have large 75% percentiles and 10 and 31 have large maxima, whereas all the others are rather small (see Fig. 9).

With pH-value, we can see that for some spots (such as 17, 18, 33) the values are rather large and for other spots (suc as 25) the values are rather small. That is, 17, 18, 33 are more alkaline, whereas 25 is more acid (remember: 7 would be neutral; see Fig. 10).

This type of plots allows to look at all observations of all spots at the same time. Unfortunately, we are not able to see the location of the spots. In DAS we have the possibility to plot a map of the locations. Moreover, we can use one certain time and plot a coded variable on these spots. For example,in Figure 11 we can see a map of the variable dust at time 13 (that is: 5.12.1985 - 2.1.1986). The special symbol in Figure 11 indicates the chimney stack. The larger the circles, the larger are the measured values. The explanation tells us how the borders of each class have been defined.

To see the development of a variable through the time we plot a map for each time through one year (see Figs. 12, 13, and 14). In Figure 14 we can see that at some periods the pH-value has not been measured, therefore only the locations are plotted.

REFERENCE

Dutter, R., Leitner, T., Reimann, C., and Wurzer, F., 1980, DAS: Data Analysis System, numerical and graphical statistical analysis, mapping of regionalized (e.g. geochemical) data on personal computer: Preliminary handbook: Technical Univ. Wien.

INDEX